30 Mathematics Lessons
Using the TI-15™

Author

Pamela Dase, M.A.Ed., Certified National T³ Instructor

Senior Editor
Sara Johnson

Associate Editor
Torrey Maloof

Assistant Editor
LeAnne Bagnall

Editorial Director
Dona Rice

Editor-in-Chief
Sharon Coan

Editorial Manager
Gisela Lee, M.A.

Creative Director
Lee Aucoin

Cover Designer
Neri Garcia

Interior Layout Designer
Robin Erickson

Print Production Manager
Don Tran

Mathematics Consultants
Lori Barker
Carolyn Barwick Belson, M. Ed.
Melanie Easterday
Karie Feldner Gladis, M.S.Ed.

Publisher
Corinne Burton, M.A.Ed.

Shell Education

5301 Oceanus Drive
Huntington Beach, CA 92649-1030
http://www.shelleducation.com
ISBN 978-1-4258-0616-3
© 2010 by Shell Educational Publishing, Inc.
Reprinted 2012

The classroom teacher may reproduce copies of materials in this book for classroom use only. The reproduction of any part for an entire school or school system is strictly prohibited. No part of this publication may be transmitted, stored, or recorded in any form without written permission from the publisher.

Table of Contents

Introduction
Research.. 5
Correlation to Standards........................... 7
How to Use This Book................................ 9
    Lesson Components................................ 9
    Integrating This Resource Into Your Mathematics Curriculum..................... 11
    Implementing the Lessons................... 11
    Instructional Time Line........................ 12
    Instructional Plan................................. 13
    Vocabulary Activities............................ 14
    Differentiating Instruction................... 22
    Grouping Students................................. 23
    English Language Learners................. 24
Overview of the TI-15................................. 26
    About the TI-15...................................... 26
    Quick Reference to Keys...................... 28
Utilizing and Managing the TI-15s in the Classroom 31
    Methods for Teaching with the TI-15 31
    Storing and Assigning the TI-15s............ 32
    Distributing the TI-15s 32
    Checking for Damage and Returning the TI-15s... 32
    Check-Off List 33
    Damage Report 34
    Facilitating a TI-15 Center 35
    Classroom Layouts.............................. 35
    TI-15 Center Rotation Schedule 35
    Parent Letter....................................... 36
Assessment... 37
    Completion Grades 37
    Using a Point System for Formal Grades 37
    Grading with a Rubric......................... 37
    Completion Grades Template 38
    General Rubric..................................... 39
    Create Your Own Rubric 40
References Cited 41

Unit 1: Number and Operations
Lesson 1: Close Enough............................. 42
    Round It Up 45
    More Money... 47
Lesson 2: Leftovers..................................... 48
    Divvy It Up!.. 51
    By the Numbers................................... 53
Lesson 3: Half Again.................................. 54
    It All Adds Up...................................... 57
    Different Halves................................... 59
Lesson 4: More or Less 60
    Fair Prices... 63
    Who Has More?.................................... 65
Lesson 5: Big Buck$ 66
    Uncle Buck$ 69
    Dollars and Days.................................. 71
Power Lesson 1: Favorites........................ 72
    Group Response Sheet 77
    Tally Sheet... 78
    Fractions and Decimals...................... 79
    The Rest of the Story 81

Unit 2: Algebra and Algebraic Thinking
Lesson 6: How Many Cookies?................. 82
    Sugar Cookies..................................... 85
    Lots of Cookies 87
Lesson 7: Fill in the Blank......................... 88
    Missing Numbers 91
    Write Your Own 93
Lesson 8: Moving Along............................ 94
    Snail Mail.. 97
    Hikers 99
Lesson 9: Order Matters........................... 100
    Which Way Did They Go? 103
    All Mixed Up 105

Table of Contents (cont.)

Lesson 10: The Shapes of Numbers 106
    Dots 109
    The Italian Connection 111
Unit 2 Power Lesson: A New Look 112
    Dozens and Dozens 117
    The Cookie Business 119

Unit 3: Geometry

Lesson 11: Mini Me 122
    Scaled Down 125
    Scaled Up 127
Lesson 12: On the Ball 128
    Big and Small 131
    Another Look 133
Lesson 13: It's on the Map 134
    Mapping Measurement Mountain 137
    My Math Map 139
Lesson 14: Scale That Structure 140
    Around the World 143
    My Structure 144
Lesson 15: Painting Flags 146
    Similar and Congruent 149
    Stripes 150
Power Lesson 3: Shape Investigations 152
    Shape Scavenger Hunt 157
    Investigating Shapes 159
    Measuring Irregular Shapes 160
    Designing a Park 161

Unit 4: Measurement

Lesson 16: Hamster Haven 162
    Harvey the Hamster 165
    Four Walls 167
Lesson 17: Measuring Mania 168
    Use Your Feet! 171
    Measure That Rectangle 172
Lesson 18: Cover It Up 174
    Beneath the Feet 177
    Kitchen Floor 179

Lesson 19: Investigating Volume 180
    Boxes of Volume 183
    Build It! 184
Lesson 20: Fish for the Science Fair 186
    Fish Frenzy 189
    Fish Food 190
Unit 4 Power Lesson: Easy as Pi 192
    Parts of a Circle 197
    Round About 198
    Close Enough? 199

Unit 5: Data Analysis and Probability

Lesson 21: In the Long Run 202
    Heads or Tails 205
    Fair Play 207
Lesson 22: Bowling for Dollars 208
    Rolling Along 211
    On a Roll 213
Lesson 23: It's in the Bag 214
    Bag of Chips 217
    Mystery Chips 219
Lesson 24: They Add Up 220
    Let's Get Rolling 223
    Toss Those Cubes! 225
Lesson 25: Typically 226
    Birthday Weather 229
    Birthday Game 231
Unit 5 Power Lesson: Graph It! 232
    Frozen Flavors 237
    A Flavorful Dessert 240
    Favorite Flavors 241
    Top That! 242

Appendices

Appendix A: Answer Key 243
Appendix B: Teacher Resource CD Contents 271

Introduction

Research

The role of the handheld calculator in the teaching and learning of mathematics has been intensely examined ever since its introduction in the 1970s. When calculators were first introduced into the classroom, they were often used to completely replace computation and were rarely used for developing concepts. The initial concern of mathematics educators was that students who used calculators would not learn basic arithmetic skills. Educators feared that if students were not skillful in performing basic arithmetic operations using traditional methods, they would be unable to learn and apply mathematical concepts. (Ellington 2003).

The mathematical education community began studying the ways that calculators could be implemented into the mathematics curriculum. Early research showed mixed results. Some studies showed that students who had used calculators were less competent in paper and pencil computations. Other studies showed little difference. A few studies looked at the mathematical understanding of students using calculators and showed that some students using calculators showed increased understanding (Ellington 2003). In 1974, the National Council of Teachers of Mathematics (NCTM) endorsed the use of the calculator in mathematics, stating that the tool "would aid algorithmic instruction, support concept development, reduce demand for memorization, enlarge the scope of problem solving, provide motivation, and encourage discovery, exploration, and creativity."

Researchers then began to investigate ways in which the use of calculators enhances student performance. Many studies found that students who had used calculators were more capable of applying mathematics in problem solving, developed better number sense, and had a greater understanding of basic concepts. Students were also found to have a more positive attitude towards mathematics (Ellington 2003). Arithmetic calculators were also found to help students work with unusual problem representations and adapt solution strategies by being able to focus on the problem instead of the computations (Ruthven and Chaplin 1997).

In 2000, NCTM recommended "the integration of calculators into the school mathematics program at all grade levels." The position was amplified with the statement: "Research and experience support the potential for appropriate use to enhance the learning and teaching of mathematics. Not only has calculator use been shown to enhance cognitive gains in the expected area of number sense, it has also been shown to improve conceptual development and visualization (NCTM 2000). In addition, it provides greater efficiency. Teaching with technology has been shown to lead to more efficient, student-centered learning, freeing time for students to concentrate on the more creative translation phase of problem solving (Kutzler 2000).

Eventually, it became clear that calculator use alone does not improve student learning. Researchers investigated how calculator use was implemented in the classroom. Studies showed that the determining factor in the efficacy of calculator use depends more on the way the classroom teachers and educational materials incorporate their use. When calculators were just used as a substitute for mental or paper and pencil computation, they had little effect on student understanding. When calculators are used to investigate patterns by enabling students to easily look at many different examples and form their own understanding, the students who use calculators exhibit greater understanding of mathematics and problem-solving skills (Reys and Arbaugh 2001).

Introduction (cont.)

Research (cont.)

The advantages of integrating the use of calculators into daily instruction are now well documented. Even if calculators are used for computation, it can be argued that as our society increasingly uses technology to make work more efficient and productive, our students can use calculators to do their work more efficiently and use the results to answer questions and influence decision making (Reys and Arbaugh 2001). The use of calculators also allows students to explore the mathematics of a problem involving computations that would make it inaccessible to most students. In this way, calculator use enables students to pursue problem solving in depth (Stiff 2001).

However, calculators are most effective when used with specific goals in mind. Using calculators as a teaching and learning tool helps students develop conceptual understanding. Students can be given situations to analyze. Using the calculator allows them to describe patterns, reason through the steps of a problem, and justify their work (Chval and Hicks 2009). NCTM emphasizes that the calculator is a tool for exploring number concepts and generating data that can be studied for patterns (NCTM 2000).

It is also true that the use of technology is present in all parts of the lives of students and their parents. By engaging with a calculator as part of mathematics learning, children are learning about and using the tools of society as well as developing a deeper understanding of mathematics. Not only is it unrealistic to teach mathematics without the use of calculators, it also does not allow students to learn to use technology effectively. Students learning with the aid of technology are becoming techno-literate and are developing number sense (Sparrow and Swan 2005).

Unfortunately, calculator use is still often viewed as an add-on to teachers' lessons. Many textbooks do include the use of calculators. However, in most of these books, calculator use is denoted as optional, or is used exclusively for computing or checking answers. The textbooks do not integrate calculator use into lessons or use it as an educational tool. Although students may learn how to use calculators for computation, the educational advantages of calculator use are overlooked (Chval and Hicks 2009).

The TI-15 is a pedagogically sound tool that helps students make connections between classroom learning and real-world situations. It helps students explore their world through investigation and experimentation, and helps develop skills in addition, subtraction, powers, and answer format. Its large display allows the students to see the expression they entered and its simplified value. It includes functions that allow students to go beyond basic computation to investigate real mathematics. The exercises in this book are designed to take advantage of these properties.

Introduction (cont.)

Correlation to Standards

The No Child Left Behind (NCLB) legislation mandates that all states adopt academic standards that identify the skills students will learn in kindergarten through grade 12. While many states had already adopted academic standards prior to NCLB, the legislation set requirements to ensure that the standards were detailed and comprehensive.

Standards are designed to focus instruction and guide adoption of curricula. Standards are statements that describe the criteria necessary for students to meet specific academic goals. They define the knowledge, skills, and content students should acquire at each level. Standards are also used to develop standardized tests to evaluate students' academic progress.

In many states today, teachers are required to demonstrate how their lessons meet state standards. State standards are used in the development of Shell Education products, so educators can be assured that they meet the academic requirements of each state.

How to Find Your State Correlations

Shell Education is committed to producing educational materials that are research and standards based. In this effort, all products are correlated to the academic standards of all 50 states, the District of Columbia, and the Department of Defense Dependent Schools. A correlation report customized for your state can be printed directly from the following website: **http://www.shelleducation.com**. If you require assistance in printing correlation reports, please contact Customer Service at 1-800-877-3450.

McREL Compendium

Shell Education uses the Mid-continent Research for Education and Learning (McREL) Compendium to create standards correlations. Each year, McREL analyzes state standards and revises the compendium. By following this procedure, they are able to produce a general compilation of national standards.

Each lesson in this book is based on one or more of the McREL content standards. To see a state-specific correlation, visit the Shell Education website at **http://www.shelleducation.com**.

NCTM Standards Correlations

Each lessons also correlates to NCTM standards. The chart on the following page shows the correlations to the National Council for Teachers of Mathematics (NCTM) standards.

Introduction (cont.)

Correlation to Standards (cont.)

NCTM Standard	Lesson Title and Page Number
Understand numbers, ways of representing numbers, relationships among numbers, and number systems	Half Again (p. 54); Big Buck$ (p. 66); Favorites (p. 72); It's in the Bag (p. 214); How Many Cookies? (p. 82)
Compute fluently and make reasonable estimates	Close Enough (p. 42); Half Again (p. 54)
Understand meanings of operations and how they relate to one another	Leftovers (p. 48); More or Less (p. 64)
Understand patterns, relations, and functions	Moving Along (p. 94); Order Matters (p. 100); The Shapes of Numbers (p. 106); A New Look (p. 112)
Represent and analyze mathematical situations and structures using algebraic symbols	Fill in the Blank (p. 88)
Use mathematical models to represent and understand quantitative relationships	Mini Me (p. 122); It's on the Map (p. 134); Scale That Structure (p. 140)
Analyze characteristics and properties of two- and three-dimensional geometric shapes and develop mathematical arguments about geometric relationships	Painting Flags (p. 146); On the Ball (p. 128); Shape Investigations (p. 152)
Understand measurable attributes of objects and the units, systems, and processes of measurement	How Many Cookies? (p. 84); Measuring Mania (p. 168); Fish for the Science Fair (p. 186)
Apply appropriate techniques, tools, and formulas to determine measurements	Hamster Haven (p. 162); Measuring Mania (p. 168); Cover It Up (p. 174); Investigating Volume (p. 180); Fish for the Science Fair (p. 186); Easy as Pi (p. 192)
Formulate questions that can be addressed with data and collect, organize, and display relevant data to answer them	Typically (p. 226); Graph It! (p. 235)
Select and use appropriate statistical methods to analyze data	Bowling for Dollars (p. 208); Typically (p. 226); Graph It! (p. 232)
Understand and apply basic concepts of probability	In the Long Run (p. 202); It's in the Bag (p. 214); They Add Up (p. 220)

Introduction (cont.)

How to Use This Book

30 Mathematics Lessons Using the TI-15 was created to provide teachers with strategies for integrating TI-15 technology into their instruction for common mathematical concepts. The lessons are designed to move students from the concrete through the abstract to real-life application, while developing students' TI-15 skills. The table below outlines the major components and purposes for each lesson.

Lesson Components
Overview and Mathematics Objective • A description of the concepts students will learn • Includes the mathematical standard for the lesson **TI-15 Functions** • TI-15 functions used in the lesson **Materials** • Lists the activity sheets included with each lesson • Lists any additional resources needed for each lesson **Vocabulary** • Lists important vocabulary terms used in the lesson • Lists a vocabulary activity to help students learn the words **Warm-up Activity** • A fun and quick introduction activity that uses the TI-15 
Explaining the Concept • Instructional strategies to accompany teacher-directed activity sheet • Concrete instructional methods for promoting students' understanding of mathematical concepts • Incorporates TI-15 technology to promote student understanding • Keystrokes to provide visual support • Detailed step-by-step instructions for teaching the lesson 
Applying the Concept • Instructional strategies to accompany student activity sheets **Differentiation** • Additional ideas for differentiation strategies within the application problems **Extending the Concept** • Additional ideas for practicing concepts and skills that can be used to review, extend, and challenge students' thinking 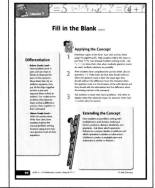

Introduction (cont.)

How to Use This Book (cont.)

Lesson Components

Activity Sheets

- Student reproducibles with easy-to-follow directions
- Student-friendly steps for using the TI-15
- Often requires students to explain their problem-solving strategies and mathematical thinking
- Provides places for student responses

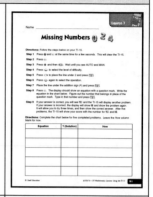

Power Lessons

- Lessons are intended to be conducted over a few days
- Lessons in which students often work together and rely on prior knowledge, skills, and experiences to solve problems and find answers
- Lessons include: overview, mathematics objective, TI-15 functions list, materials list, vocabulary, warm-up activity, multiple Explaining the Concept and Applying the Concept sections broken down into parts, and multiple activity sheets

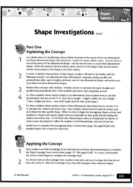

Teacher Resource CD

- Contains copies of all student activity sheets
- Teacher resource materials referred to in the introduction
- Student reproducibles such as graphs, grids, and a hundred chart
- Student-friendly glossary of all vocabulary words used in the lessons
- Full-color image of the TI-15

Border Key

- The border along the top of each lesson page indicates the mathematics strand covered in that lesson. Please use the key below as a reference.

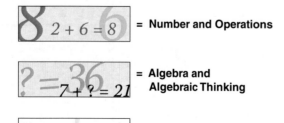 = Number and Operations

= Algebra and Algebraic Thinking

= Measurement

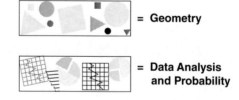 = Geometry

= Data Analysis and Probability

Introduction (cont.)

How to Use This Book (cont.)

Integrating This Resource Into Your Mathematics Curriculum

When planning instruction with this resource, it is important to look ahead at your instructional time line and daily lesson plans to see where *30 Mathematics Lessons Using the TI-15* can best be integrated into your curriculum. As with most lessons, the majority of the planning takes place before the students arrive.

Each lesson begins with an overview. Preview this description to find a lesson that correlates with the objective listed in your time line. The *Instructional Time Line* template (page 12) is provided to help integrate this resource into long-range planning.

Implementing the Lessons

After integrating this resource into your instructional time line, use the steps below to help you implement the lessons. The *Instructional Plan* template (page 13) is provided to help determine the resources and lessons to be used for the instructional phases: Explaining the Concept, Applying the Concept, Differentiation, and Extending the Concept.

1. Familiarize yourself with the lesson plan. Make sure you have all the materials needed for the lesson.

2. Determine how you want to pace the selected lesson. Each of the lesson parts or instructional phases are mini-lessons that can be taught independently or together, depending on the amount of instructional time and the students' needs.

   - For example, you may choose to use the Explaining the Concept section in place of the lesson taught in the textbook, or use the Applying the Concept and Extending the Concept sections to supplement the textbook.

   - The lesson parts can be taught each day for two or three days, or the lesson can be modified and all three parts can be taught in the course of a 50- to 90-minute instructional block.

3. Solve the problems before class to become familiar with the features on the TI-15, as well as with the mathematical concepts presented.

4. Because space is limited in lesson plan books, use a three-ring binder or a folder to keep detailed plans and activities for a specific concept together.

Introduction (cont.)

How to Use This Book (cont.)

Directions: In the first column, record the date or the days. In the second column, record the standards and/or objectives to be taught that day. In the third and fourth columns, write the lesson resources to be used to teach that standard and the specific page numbers. In the fifth column, include any adaptations or notes regarding the lesson resources.

Instructional Time Line

Date	Standards/ Objectives	Lesson Resources (e.g., *30 Mathematics Lessons Using the TI-15*, textbook)	Pages	Adaptations or Notes

Introduction (cont.)

How to Use This Book (cont.)

Directions: In the first column, write the date(s) of the lesson. In the second column, write the standards and/or objectives to be taught. In the remaining columns, write the lesson resources and page numbers to be used for each phase of instruction, as well as any notes and plans for modifying the lessons or differentiating instruction.

Instructional Plan

Date	Standards/ Objectives	Explaining the Concept	Applying the Concept	Extending the Concept	Assessments	Adaptations/ Differentiation
		pgs.	pgs.	pgs.	pgs.	
		pgs.	pgs.	pgs.	pgs.	
		pgs.	pgs.	pgs.	pgs.	
		pgs.	pgs.	pgs.	pgs.	
		pgs.	pgs.	pgs.	pgs.	

Lesson Resources Per Instructional Phase

Introduction (cont.)

How to Use This Book (cont.)

Vocabulary Activities

Vocabulary knowledge has been proven to be a key component of reading comprehension, as well as being strongly related to general academic achievement (Feldman and Kinsella 2005). Similarly, vocabulary deficiencies have been linked to academic failure among students (Becker 1977, as cited in Baker, Simmons, and Kame'enui 1995). Students need to understand key vocabulary words specific to the mathematics topic they are studying in order to fully comprehend the concepts.

Some educators differentiate between general academic vocabulary and specialized content vocabulary when discussing academic vocabulary. What is the difference? General academic vocabulary includes high-utility words found across content areas. These words are those that students will likely find throughout their academic reading and writing experiences and use in academic speech. Words such as *features*, *attributes*, *principle*, *perspective*, *compatible*, and *influence* are examples of general academic vocabulary. Specialized content vocabulary includes words that are specific to a particular content area and represent important concepts or ideas for students to learn. Examples of specialized content vocabulary include *democracy*, *forensic*, *numerator*, and *protagonist*. In this book, all specialized content vocabulary is related to mathematical concepts that are covered in each lesson. In each lesson, students will learn specialized content words such as *place value*, *sum*, *addend*, and *inverse*. Specialized content vocabulary is considered a part of academic vocabulary.

The National Reading Panel Report (2000) identified academic vocabulary as essential in the development of students' reading skills. Academic vocabulary includes terms that are used throughout schooling. English teacher and author, Jim Burke (no date) identified almost 400 academic vocabulary words that students must know by the time they enter middle school. Terms such as *solve*, *quotient*, and *sum* are commonly understood by proficient students, but those who only have a limited understanding of these terms may not be able to complete the simplest mathematical problems. Further, a variety of teaching strategies is necessary to teach academic vocabulary. The National Reading Panel Report (2000) found that vocabulary is learned both indirectly and directly, and that dependence on only one instructional method does not result in optimal vocabulary growth.

Content-area vocabulary is also highly specialized with words that are not typically used in everyday life. Therefore, all students need explicit introduction to those words to understand the text. The task is even more complicated for English language learners and struggling readers. According to Feldman and Kinsella (2005), "Developing readers cannot be expected to simply 'pick up' substantial vocabulary knowledge exclusively through reading exposure without guidance. Specifically, teachers must design tasks that will increase the effectiveness of vocabulary learning through reading practice."

While students' vocabulary skills may not be the primary concern in mathematics, the literature certainly suggests it must be an important consideration. It is not enough to give students a list of words and have them look up the definitions in dictionaries or glossaries. Students who are struggling with vocabulary are not going to find the process easier by simply being given more words to sort through (Echevarria, Vogt, and Short 2004). Struggling readers and English language learners need context-embedded activities that acquaint them with the necessary and most central words for comprehension of the content. This is why it is so important to use a variety of strategies to teach mathematical vocabulary.

Introduction (cont.)

How to Use This Book (cont.)

Vocabulary Activities (cont.)

Explicit instruction of key words involves a much more thorough and well-planned instruction of critical terms. Feldman and Kinsella (2005) advocate a consistent, five-step approach to teaching students new words.

1. **Pronounce**—A teacher must share correct pronunciations of vocabulary, but should not be the only person in the room who talks. Students need to repeat and pronounce new words. Call on students to say the word together several times. Encourage them to turn to partners and repeat the word.

2. **Explain**—Explaining a new vocabulary word does not mean stating its dictionary definition. Students need to hear a simple and accessible explanation of the new word using student-friendly language. Sharing a familiar synonym or a connected word is also useful in teaching new terms.

3. **Provide Examples**—Once students hear what a word means, they still need to think about how the word definition fits with what they already know. Students often require two or three examples to help them understand a word's true meaning. Examples should be familiar to students and, when possible, relevant to their lives and experiences. The examples should also be varied so that students begin to understand the different ways that a word can be used.

4. **Elaborate**—After hearing explanations and examples, students need to interact with new words in more personal ways. This may mean that students share their own examples of a word. They might visualize something related to the word that helps them remember its meaning. They might use the word in a way that makes sense in their own lives. All of these suggestions will help students think about vocabulary and how it fits within their knowledge base.

5. **Assess**—It is important to incorporate assessments into your vocabulary instruction. These assessments may range from quick, informal checking for understanding to a lengthier, more formal test or quiz.

The true best practices for teaching vocabulary are those that spark curiosity and motivate students to learn more about new words. Historically, vocabulary instruction in the classroom has often been characterized by rote memorization, meaningless dictionary work, and tedious oral or written assignments. These kinds of activities do not create excited learners who are eager to hear, speak, and read new words. Instead, the norms for such learning must be shifted so that instruction is rich and dynamic and is focused on developing genuine interest in words. Word learning should not be confined to the classroom, only to occur at those times of the day set aside for vocabulary. Students should be encouraged to always look and listen for interesting and unique words to learn more about and share. When students are able to use their skills at learning unfamiliar vocabulary throughout their daily lives outside of school, teachers can sit back and watch these word detectives go to work!

The vocabulary activities on pages 16–21 offer teachers effective word-learning strategies that support vocabulary knowledge and conceptual understanding in mathematics.

Introduction (cont.)

How to Use This Book (cont.)

Vocabulary Activities (cont.)

Sharing Mathematics

1. Write the vocabulary words specific to the day's lesson on the board or overhead and give three to four counters/markers to each student.

2. Place students into small groups of three to five.

3. Tell students the following rules:

   - Every time a student says a sentence with one of the vocabulary words in it, he or she places a marker in a cup, bowl, or small paper bag.

   - The sentences must be definitions or example sentences. You may want to share good examples (e.g., My favorite decimal is 0.75 because Grandpa is 75.) and weak examples (e.g., I like fractions.).

   - The students have to use all of their markers.

   - Once a student's markers are "spent," he or she is not allowed to say anything more.

   - Students need to be respectful and listen to the vocabulary sentences from all their classmates.

4. The students generate sentences using the lesson vocabulary words until all students have "spent" their markers.

Sentence Frames

1. Write the vocabulary words specific to the day's lesson on the board or overhead.

2. Share a simple sentence frame that presents a vocabulary word in the proper mathematical context. The sentence frame has blanks in which the students will substitute information. You should write the sentence frame on the board or overhead.

   - This example uses the sample vocabulary words *equation* and *equivalent*.

     The equation _____ is equivalent to the equation _____.

3. Write sample answers to put in the blanks.

   - For this example, the teacher writes: $3 + 5 = 8$; $9 - 1 = 8$; $6 - 2 = 4$; $4 + 0 = 4$.

4. Have students practice reading the sentence using the sample answers.

5. Have students work in pairs to write their own sentence frames.

6. Have a few pairs of students share the sentence frames they created.

Introduction (cont.)

How to Use This Book (cont.)

Vocabulary Activities (cont.)

Which Statement Is Accurate?

In this activity, each vocabulary word specific to the day's lesson is used in four written sentences. You can write these ahead of time. For each vocabulary word, three accurate mathematical sentences are written and one inaccurate mathematical sentence is written.

> Sample sentences using the vocabulary word *fraction*:
>
> 1. The number $\frac{1}{11}$ is a fraction.
> 2. The number $\frac{3}{4}$ is a fraction.
> 3. The number 65 is a fraction.
> 4. The number $\frac{7}{5}$ is a fraction.

1. Write the vocabulary words specific to the day's lesson on the board or overhead.

2. Display four sentences on the board or overhead. Three are accurate and one is inaccurate. The sentences are numbered one to four.

3. The students work in groups of four. Each group reads the sentences aloud.

4. Individual students decide which sentence is inaccurate. They each hold the number with their fingers while hiding it from their group (e.g., holding three fingers up—but hidden from the group—if number 3 is the inaccurate sentence).

5. When everyone has decided, each individual shows his or her guess to the team. The team discusses the answer to reach a consensus. When the team has reached a consensus, one student writes the answer on a small piece of paper.

6. When all teams have reached a consensus, each team displays its answer to the rest of the class. The class can discuss each team's results.

7. If time permits, the teams can convert the inaccurate sentence to an accurate sentence by making appropriate changes.

Introduction (cont.)

How to Use This Book (cont.)

Vocabulary Activities (cont.)

Total Physical Response (TPR)

1. Write the vocabulary words specific to the day's lesson on the board or overhead.

2. Play Simon Says where "Simon" says a concept or vocabulary word and students perform an action that goes with that word. For example, Simon says "triangles," and students form triangles with their bodies.

Gouin Series

1. Write the vocabulary words specific to the day's lesson on the board or overhead.

2. Prepare a series of short statements describing a logical sequence of actions related to a concept. These statements should include concrete action verbs using the same tense and person throughout.

   Example:

   Finding the Missing Angle in a Triangle

   - *I look for two given angles' measures.*
   - *I add the two given angles.*
   - *I subtract the sum of the angles from 180°.*
   - *I found the missing angle measure.*

3. Discuss the language structures in the statements and chorally say them as you act them out. It may be useful to have props and visuals to support student understanding of the statements.

Music Makers

1. Write the vocabulary words specific to the day's lesson on the board or overhead.

2. Have students make up or sing songs about learning concepts. This is a fun way to use content language in context. When students are making up the songs, be sure they are using the language accurately rather than in nonsensical ways.

   Example:

   Sing to the tune of "Head, Shoulders, Knees, and Toes."

   > *Squares have four equal sides!*
   > *Equal sides!*

Introduction (cont.)

How to Use This Book (cont.)

Vocabulary Activities (cont.)

Chart and Match

1. Write the vocabulary words specific to the day's lesson on the board or overhead.

2. Students will create a three-column grid labeled with the following headings: *Word*; *Illustration or Example*; and *Definition or Description*. Students write the vocabulary words down the left side of the grid. One word goes in each row.

3. Introduce the words and lead a whole-class discussion. Use examples and draw pictures to show the vocabulary words.

4. Direct students to draw pictures or to write examples of each vocabulary word in the middle column of the grid, next to the corresponding vocabulary word.

5. In the final column, the class decides on a way to describe or define each word. This should not be a dictionary definition. Rather, after the discussion, the students can write their understandings of the word, or you can help them write student-friendly explanations.

Vocabulary Bingo

1. Write the vocabulary words specific to the day's lesson on the board or overhead and give each student nine counters/markers.

2. The students need to create a 3 x 3 grid, like a bingo board. They should write one word in each square. They should write mathematical vocabulary words from previous lessons in empty squares. (Creating these charts ahead of time would save time during class.)

3. The students are encouraged to draw small pictures or examples next to the location where they placed each word on their bingo charts if there are any words they already know.

4. Lead a class discussion of the meanings and examples of the day's vocabulary words. Students should draw examples to help them understand the words they are not familiar with.

5. Start the bingo learning activity. Read a description of one vocabulary word or show an example or representative picture. Students will locate the word and cover it with a marker. Cut-up pieces of paper or small manipulatives can be used. If students wrote the same vocabulary word twice, they can cover both spaces. Play ends when a student correctly covers a row, column, or diagonal with markers.

Introduction (cont.)

How to Use This Book (cont.)

Vocabulary Activities (cont.)

Frayer Model

The Frayer Model (Frayer, Frederick, and Klausmeier 1969) is a strategy in which students use the graphic organizer as a means to better understand a concept and to distinguish that concept from others they may know or may be learning. The framework of the Frayer Model includes the concept word, the definition, the characteristics of the concept word, examples of the concept word, and non-examples of the concept word. It is important to include both examples and non-examples so that students are able to clarify what the concept word is and what it is not. The Frayer Model is especially useful for teaching mathematics vocabulary that describes complex concepts or vocabulary that describes mathematical concepts students may already know but cannot yet clearly define.

Instruct students to write down the word for a new concept they are learning on the Frayer Model graphic organizer found on the Teacher Resource CD (frayer.doc). First, the teacher and students must define the concept and list its attributes.

Next, distinguish between the concept and similar concepts with which it might be easily confused. When doing this, help the students to understand the concept in depth. This can easily be accomplished through question and answer during a short discussion. Also, provide students with examples of the concept and explain why they are examples. Next, provide the students with non-examples of the concept and explain why they are non-examples. Discuss the examples and non-examples at length. Encourage students to generate their own examples and non-examples, and allow them to discuss their findings with the class. Once students are skilled at using the strategy, the entire class can work in pairs to complete a Frayer Model graphic organizer for different essential concepts and then present their findings to the class.

A sample of the Frayer Model graphic organizer is provided below.

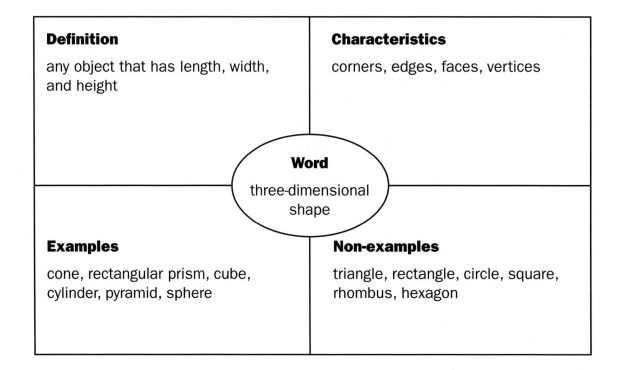

Introduction (cont.)

How to Use This Book (cont.)

Vocabulary Activities (cont.)

Word Wizard

The Word Wizard strategy (Beck, McKeown, and Kucan 2002) extends student learning beyond the school day by rewarding students for using and noticing mathematics vocabulary words in their lives. Teachers create a Word Wizard chart with each student's name and room for tally marks, stickers, or some kind of marker. As students bring in evidence of word use, they are rewarded with marks or stickers on the chart.

The Word Wizard strategy should be used after students have developed a good understanding of the selected vocabulary words (e.g., at the middle or end of a unit). Encourage students to pay attention to words and notice when they see, hear, or use focus vocabulary words beyond the context of classroom learning. This helps students connect these words to their personal experiences.

A Word Wizard Chart is located on the Teacher Resource CD (wizard.doc).

1. Use the Word Wizard resource page to create a Word Wizard chart that includes each student's name. There should be room for you to make tally marks, add stickers, or include some other marker when students report back about vocabulary words they used or noticed outside of the classroom.

2. Share the chart with students. Explain that you want to record any time they see, hear, or use a vocabulary word outside the classroom. As a group, discuss how this may occur in their lives. Students will report where they noticed the word and how it was used.

3. It is also fun to create a Word Wizard Hat. If a student uses or recognizes two or three of the vocabulary words, he or she is rewarded with a hat displaying the words. Worn around school, the hat indicates to other adults that they should ask the student about the word definitions. This creates opportunities for students to share knowledge and for adults to heap on praise!

4. As students return to class with evidence of word use, set aside time to discuss their examples. Ask each student to share his or her Word Wizard moment. This should include discussion of the vocabulary word, as well as how it was used or where it was noticed. Allow some time for whole-class discussion, but also give students time to describe their experiences in pairs.

   The teacher should take time during the day to mark the chart and have students share their experiences. Each day the teacher gives them time to discuss their experiences in pairs. After a few days, the teacher reviews the chart with students and leads a group discussion of using and/or noticing vocabulary beyond the classroom.

5. In conclusion, ask students how acting as Word Wizards helps them to be more aware of new vocabulary words.

Introduction (cont.)

How to Use This Book (cont.)

Differentiating Instruction

Students in today's classrooms have a diverse range of ability levels and needs. A teacher is expected to plan and implement instruction to accommodate gifted students, and above-level, on-level, and below-level students. The lessons in this resource can be differentiated by their content (what is taught), process (how it is taught), and product (what is created). Below are some strategies that can be used to adapt the lessons in this resource to meet most students' needs. This is not an all-inclusive list, and many of the strategies are interchangeable. It is important to implement strategies based on students' learning styles, readiness, and interests.

Below Level	On Level	Above Level/Gifted
• Reduce the number of problems in a set. • Write hints or strategies by specific problems. • Simplify the text on activity sheets. • Create *PowerPoint*™ presentations of lessons and have students use them as a review or reference. • Have students take notes. • Use visual aids and actions to represent concepts and steps of a process. • Act out problems. • Model skills and problems in a step-by-step manner. • Use manipulatives to explain concepts and allow students to use them to complete assignments. • Have students work in homogeneous or heterogeneous groups. • Have students draw pictures of how they solved the problems.	• Have students take notes. • Use activities centered on students' interests. • Have students generate data. Engage students using *PowerPoint*™, games, and applets. • Have students work in homogeneous or heterogeneous groups. • Have students write explanations for how they solved the problems. • Use the Extending the Concept section to review concepts or skills.	• Have students create how-to guides for functions on the TI-15. • Have students use multimedia, such as *PowerPoint*™ to present how they solved the problems and/or used the calculator. • Have students work in homogeneous groups. • In addition to or in place of an activity sheet, assign the Extending the Concept sections. • Have students take on the role of teacher or mentor. • Have students create games for practicing concepts and skills.

Introduction (cont.)

How to Use This Book (cont.)

Grouping Students

Recommendations for cooperative groups and independent work are given throughout the lessons in this resource. The table below lists the different types of groups, a description of each, and management tips for working with each.

Group Type	Description of Group	Management Tips
Heterogeneous Cooperative Groups	Three to six students with varied ability levels	Give each student a role that suits his or her strengths. Give each group a sheet with directions for the task and a description of each role in completing that task.
Homogeneous Cooperative Groups	Three to six students with similar ability levels	Give each student an equal role in the task by having him or her take the lead in a different part or problem of each assignment.
Paired Learning	Two students with similar abilities or mixed abilities	When completing assignments, have students sit side by side and give students opportunities to manipulate the materials.
Independent Work	Students work individually	Closely monitor students' work to correct any misconceptions; this will help students retain the information and develop confidence in their abilities. This is also a good time to work one-on-one with struggling students or gifted students.

Introduction (cont.)

How to Use This Book (cont.)

English Language Learners

The following suggestions are specifically designed to help teachers differentiate mathematics for English language learners. Three strategies for doing so include the use of vocabulary activities, manipulatives, and the TPR (Total Physical Response) strategy.

Vocabulary

Teaching vocabulary in mathematics is also especially important for English language learners. Too often, students do not have an accurate understanding of mathematical terms. Researchers have identified several barriers that inhibit their learning as well (Thompson and Rubenstein 2007). Many vocabulary words have different meanings in other content areas. For example, the term *solution* has different meanings in mathematics and science. Other terms are also commonly used in everyday language, but they have more precise meanings in mathematics. Still other words are only used in mathematics. These terms must be explicitly taught to students. Vocabulary words, such as *quotient*, that are unique to the subject are difficult to learn. In this case, students are essentially learning a new language, and they need instructional strategies that will help them become aware of the new terms and apply them to problem-solving situations. (See pages 16–21 for vocabulary activities.)

Manipulatives

Manipulatives include almost any physical object used to represent an abstract concept. They are used to help students physically manipulate and act out mathematics algorithms. Manipulatives include items such as counters, base-ten blocks, pattern blocks, calculators, and even students' own fingers. The benefits of using manipulatives in the mathematics classroom are well-established. Manipulatives provide concrete materials for students to understand concepts before making abstract connections.

Total Physical Response (TPR)

TPR is a method of instruction that allows students to use physical movements as a way to acquire language skills. When utilizing this strategy, students apply actions with oral language to concepts and procedures. Teachers can have students perform the action while chorally saying the word related to it.

Proficiency Levels for English Language Learners

All teachers should know the levels of language acquisition for each of their English language learners. Knowing these levels will help to plan instruction. (The category titles and number of levels vary from district to district or state to state, but the general descriptions are common.) Students at level 1 need a lot of language support in all activities, especially during instruction. Using visuals to support oral and written language will help make the language became more comprehensible. These students "often understand much more than they are able to express" (Herrell and Jordan 2004). It is the teacher's job to move them from just listening to language to expressing language. Students at levels 2 and 3 benefit from pair work in speaking tasks, but they will need additional individual support during writing and reading tasks. Students at levels 4 and 5 (or 6, in some cases) may appear to be fully proficient in the English language. Because they are English language learners, they may still struggle with comprehending the academic language used during instruction, as well as with reading and writing.

The following chart shows a quick glance at the proficiency levels for English language learners. These levels are based on the World-Class Instructional Design and Assessment (WIDA) Consortium.

Introduction (cont.)

How to Use This Book (cont.)

English Language Learners (cont.)

Proficiency Level	Questions to Ask	Activities/Actions		
Level 1—Entering minimal comprehension no verbal production	Where is . . . ? What is the main idea? What examples do you see? What are the parts of . . . ? What would happen if . . . ? What is your opinion?	listen	draw	mime
		point	circle	
Level 2—Beginning limited comprehension short spoken phrases	Can you list three . . . ? What facts or ideas show . . . ? What do the facts mean? How is ____ related to ____? Can you invent . . . ? Would it be better if . . . ?	move	select	act/act out
		match	choose	
Level 3—Developing increased comprehension simple sentences	How did ____ happen? Which is your best answer? What questions would you ask about . . . ? Why do you think . . . ? If you could ____, what would you do? How would you prove . . . ?	name	list	respond (with 1–2 words)
		label	categorize	group
		tell/say		
Level 4—Expanding very good comprehension few errors in speech	How would you show . . . ? How would you summarize . . . ? What would result if . . . ? What is the relationship between . . . ? What is an alternative to . . . ? Why is this important?	recall	retell	define
		compare/contrast	explain	summarize
		describe	role-play	restate
Level 5—Bridging comprehension comparable to native English speakers speaks using complex sentences	How would you describe . . . ? What is meant by . . . ? How would you use . . . ? What ideas justify . . . ? What is an original way to show . . . ? Why is it better that . . . ?	analyze	defend	complete
		evaluate	justify	support
		create	express	

Introduction (cont.)

Overview of the TI-15

About the TI-15

Two-Line Display

The first line displays an entry of up to 11 characters. Entries begin on the top left. If the entry does not fit on the first line, it wraps to the second line. When space permits, both the entry and the result appear on the first line.

The second line displays up to 11 characters. If both entry and result do not fit on the first line, the result is displayed right-justified on the second line. Results longer than 10 digits are displayed in scientific notation.

If the entry does not fit on two lines, it continues to wrap, so that you always see the last two lines of the entry. You can view the beginning of the entry by scrolling up. In this case, only the result appears when you press [Enter].

Order of Operations

The TI-15 uses the following order of operations:

Priority	Functions
First	Expressions inside parentheses
Second	Determination of display of fraction results
Third	Power, square root
Fourth	Negation
Fifth	Multiplication, implied multiplication, division
Sixth	Addition and subtraction
Seventh	Conversion of a mixed number to an improper fraction or an improper fraction to a mixed number; conversion of a fraction to a decimal or a decimal to a fraction
Eighth	Completes all operations

Note: Because operations inside parentheses are performed first, you can use parentheses to change the order of operations and, therefore, change the result.

Menus

Two keys on the TI-15 display menus: [Mode] and [Frac].

Press ▲ or ▼ to move up or down through the menu list. Press ◄ or ► to move the cursor and underline a menu item. To return to the previous screen without selecting the item, press [Clear]. To select a menu item, press [Enter] while the item is underlined.

Previous Entries (History)

After an expression is evaluated, use ▲ and ▼ to scroll through previous entries and results, which are stored in the TI-15 history.

Introduction (cont.)

Overview of the TI-15 (cont.)

About the TI-15 (cont.)

Problem Solving (◈)

The Problem Solving tool has three features that students can use to challenge themselves with basic math operations or place value.

Problem Solving (Auto Mode) provides a set of electronic exercises to challenge the student's skills in addition, subtraction, multiplication, and division. Students can select mode, level of difficulty, and type of operation.

Problem Solving (Manual Mode) lets students compose their own problems, which may include missing elements or inequalities.

Problem Solving (Place Value) lets students display the place value of a specific digit, or display the number of ones, tens, hundreds, thousands, tenths, hundredths, or thousandths in a given number.

Resetting the TI-15

Pressing (On) and (Clear) simultaneously or pressing (Mode), selecting **RESET**, selecting **Y** (yes), and then pressing [Enter] resets the calculator.

Resetting the calculator:

- Returns settings to their defaults: Standard notation (floating decimal), mixed numbers, manual simplification, Problem Solving Auto mode, and Difficulty Level 1 (addition) in Problem Solving.
- Clears pending operations, entries in history, and constants (stored operations).

Display Indicators

◈	The TI-15 is in the Problem Solving feature.
■.	The TI-15 is in the Place Value feature or rounding to a specified place.
Fix	The TI-15 is rounding to a specified place.
M	Indicates that a value other than zero is stored in the memory.
→ M	Value is being stored to memory. Press [+], [−], [×], [÷], or [=] to complete the process.
Op1 or Op2	An operator and a value is stored in Op1 or Op2.
Auto	Indicates that the Problem Solving feature is in AUTO mode.
I	Integer division function has been selected (appears only when cursor is over ÷).
n/d ÷	Division results will be displayed as fractions.
N/d→n/d	The fraction result can be further simplified.
↑ or ↓	Previous entries are stored in history, or more menus are available. Press ◈ to access history. Press ◈ to access additional menu lists.
← or →	Press ◈ or ◈ to scroll and select from a menu. Press [=] to complete the selection.

Introduction (cont.)

Overview of the TI-15 (cont.)

About the TI-15 (cont.)

Error Messages

Arith Error	Arithmetical error. An invalid entry or an invalid parameter was entered.
Syn Error	Syntax error. An invalid or incorrect equation was entered.
÷ 0 Error	Divide by 0 error. An attempt was made to divide by zero.
Mem Error	Error in attempting to store an entry in memory.
Op Error	Error following steps for using Op1 or Op2.
Overflow Error	Overflow. The result is too large to fit within the boundaries of the display.
Underflow Error	Underflow. The result is too small to fit within the boundaries of the display.

Quick Reference to Keys

Keys	Description
On Clear	To reset the calculator, hold down On and Clear simultaneously for a few seconds and release. MEM CLEARED shows on the display. This will completely clear the calculator, including all mode menu settings, all previous entries in history, all values in memory, and the display. All default settings will be restored.
← →	Moves the cursor left and right, respectively, so you can scroll the entry line or select a menu item.
↑ ↓	Moves the cursor up and down, respectively, so you can see previous entries and results or access menu lists.
⌫	Deletes the character to the left of the cursor before Enter is pressed.
Mode	Displays the menu to let you select the format for results of division. Choose decimal (.) or fraction (n/d).
Mode →	Displays the menu to let you show (+1) or hide (?) the constant values in constant 1 (Op1) and constant 2 (Op2).
Mode → →	Displays the menu to allow you to clear constant values in constant 1 (Op1) or constant 2 (Op2).
Mode → → →	Displays the menu to allow you to not reset (N) or reset (Y) the TI-15.
◈	Provides a set of electronic flash cards to challenge your skills in addition, subtraction, multiplication, and division.
◈ Mode	Displays the menu to allow you to select automatic (AUTO) or manual (MAN) mode.
◈ Mode Auto →	In automatic mode, displays the menu to allow you to select the level of difficulty (1, 2, or 3).
◈ Mode Auto → →	In automatic mode, displays the menu to allow you to select the type of operation (+, −, ×, ÷, or ?).
◈ Mode HHT →	In manual mode, displays the menu to allow you to select the display options for the Place Value feature.
?	While in ◈ (Problem Solving) manual (MAN) mode, lets you indicate a missing element in an equation.

Introduction (cont.)

Overview of the TI-15 (cont.)

Quick Reference to Keys (cont.)

Key	Description
⟨⟩	While in ◈ (Problem Solving) manual (MAN) mode, lets you test inequalities. Press once to enter the less than sign (<). Press twice to enter the greater than sign (>).
■.	While in ◈ (Problem Solving), you can determine the place value of a particular digit of a given number or, in conjunction with place value keys, you can determine how many thousands, hundreds, etc., a number contains or what digit is in a given place.
■., d	Determines the place value of a digit d (0–9) of a given number.
■. 1000.	Tells how many thousands a given number contains or what digit is in the thousands place.
■. 100.	Tells how many hundreds a given number contains or what digit is in the hundreds place.
■. 10.	Tells how many tens a given number contains or what digit is in the tens place.
■. 1.	Tells how many ones a given number contains or what digit is in the ones place.
■. 0.1	Tells how many tenths a given number contains or what digit is in the tenths place.
■. 0.01	Tells how many hundredths a given number contains or what digit is in the hundredths place.
■. 0.001	Tells how many thousandths a given number contains or what digit is in the thousandths place.
Fix 1000.	Rounds results to the nearest thousand.
Fix 100.	Rounds results to the nearest hundred.
Fix 10.	Rounds results to the nearest ten.
Fix 1.	Rounds results to the nearest one.
Fix 0.1	Rounds results to the nearest tenth.
Fix 0.01	Rounds results to the nearest hundredth.
Fix 0.001	Rounds results to the nearest thousandth.
Fix ·	Removes fixed-decimal setting and returns to floating decimal.
Enter	Completes operations. Enters the equal sign or tests a solution in Problem Solving.
(−)	Enters a negative sign. Does not act as an operator.
(	Opens a parenthetical expression.
)	Closes a parenthetical expression.
Int÷	When you divide a positive whole number by a positive whole number using Int÷, the result is displayed in the form Q r R, where Q is the quotient and R is the remainder. If you use the result of integer division in subsequent calculation, only the quotient is used; the remainder is dropped.
n	When pressed after entering a number, designates the numerator of a fraction. The numerator must be an integer. To negate a fraction, press (−) before entering n.
d	When pressed after entering a number, designates the denominator of a fraction. The denominator of a fraction must be a positive integer in the range 1 through 1,000. If you perform a calculation with a fraction having a denominator greater than 1,000, or if the results of a calculation yield a denominator greater than 1000, the TI-15 will convert and display the results in decimal format.
Unit	Separates a whole number from the fraction in a mixed number.

Introduction (cont.)

Overview of the TI-15 (cont.)

Quick Reference to Keys (cont.)

Key	Description
Frac	Displays a menu of settings that determine how fraction results are displayed. • U n/d (default) displays mixed number results. • n/d displays results as a simple (improper) fraction.
Frac ⇌	Displays a menu to select the method of simplifying fractions: • MAN (default) allows you to simplify manually (step-by-step). • AUTO automatically simplifies fraction results to lowest terms.
Simp	Enables you to simplify a fraction.
Fac	Displays the factor that was used to simplify a fraction.
U$\frac{n}{d}$↔$\frac{n}{d}$	Converts a mixed number to an improper fraction or an improper fraction to a mixed number.
F↔D	Converts a fraction to a decimal, or converts a decimal to a fraction, if possible. Converts π to a decimal value.
▶%	Converts a decimal or a fraction to a percent.
√	Calculates the square root of a number.
^	Raises a number to the power you specify.
π	Enters the value of π. It is stored internally to 13 decimal places (3.1415926535897). In some cases, results display with symbolic π, and in other cases as a numeric value.
▶M	Stores the displayed value for later use. If there is already a value in memory, the new one will replace it. When memory contains a value other than 0, M displays on the screen. (The function will not work while a calculation is in process.)
MR/MC	Recalls the memory value for use in a calculation when pressed once. When pressed twice, clears memory.
Op1 Op2	Each can store one or more operations with constant value(s), which can be repeated by pressing only one key, as many times as desired.

Introduction (cont.)

Utilizing and Managing the TI-15s in the Classroom

Methods for Teaching with the TI-15

Unlike a regular calculator, the TI-15 has many functions that can be used to complete the activities in this resource. To help students feel comfortable using the TI-15, follow the steps below before starting a lesson. Be sure to allow students adequate time to explore the TI-15 before beginning the lessons.

1. Before beginning the lesson, discuss with students how to handle the TI-15 with care. Discuss such things as not slamming the calculator on their desks or pushing on the display screen, which can cause it to crack.

2. Also before beginning the lesson, demonstrate the calculator skills that students must know to be successful in the lesson.

3. To teach a skill, have students locate the keys and functions on the calculators.

4. Instruct students to press buttons slowly, especially if they have to input a series of commands or numbers. Pushing the buttons too quickly may not register all the information.

5. Continually remind students how to fix mistakes on their TI-15. Remind students they can press (On) and (Clear) simultaneously to reset their TI-15s. Resetting the TI-15 will allow students to exit any function and clear the memory of any stored numbers or constant functions. Remind students that they can also press ← to go back and fix a number or operation that was entered incorrectly, rather than having to clear the entire display.

6. If multiple steps are needed to complete the activity, list the steps on the board or on the overhead for the students to use as a reference while working.

7. Ask students who are familiar with and comfortable using the TI-15 to assist others. Let the other students know who those TI-15 mentors are.

8. Allow time to address any questions that the students may have after each step or before continuing to the next part of the lesson.

Introduction (cont.)

Utilizing and Managing the TI-15s in the Classroom (cont.)

Storing and Assigning the TI-15s

It is important to establish a management system that works best for you and your students. Here are a few suggestions:

- Before using the TI-15s with students, number each calculator with a permanent marker or label.

- Assign each student or pair of students a calculator number. Since each student will be using the same calculator every time they are distributed, this number will help keep track of any TI-15s that may be damaged or lost.

- Store the TI-15s in a plastic shoebox or an over-the-door shoe rack. Number the pockets on the shoe rack with the same numbers as the calculators.

Distributing the TI-15s

To distribute the calculators, consider at which point during the class period students will need to use them.

- If students will need their TI-15s at the beginning of the class period, write *Get your calculators* on the board or overhead.

- If the TI-15s are stored in the plastic container, make sure they are in numerical order. This will help students find their TI-15s faster.

- If the TI-15s are stored in an over-the-door shoe rack, a note to take a calculator can be placed on the door. This way, the students can grab their TI-15s as they walk into the classroom. If you use an over-the-door shoe rack, be sure to safely secure it to the door.

- If the TI-15s will not be used until later in the class period, have students retrieve their calculators by rows.

- Once the students have their TI-15s, use the *Check-Off List* (page 33) to keep track of which calculators have been used during that day or class period.

Checking for Damage and Returning the TI-15s

After distributing the TI-15s, have students check their calculators for any damage.

- If a calculator is damaged, complete the *Damage Report* (page 34). You can then refer back to the *Check-Off List* (page 33) to see which student last used the calculator.

- Have students return their TI-15s by rows.

- If the TI-15s are stored in a plastic shoebox, have them put the calculators back in numerical order.

- If the TI-15s are stored in an over-the-door shoe rack, have the students place them in the correct pockets.

- DO NOT forget to count the TI-15s before students leave.

Introduction (cont.)

Utilizing and Managing the TI-15s in the Classroom (cont.)

Check-Off List								
		DATE						
Student Name	TI-15 Number							

Introduction (cont.)

Utilizing and Managing the TI-15s in the Classroom (cont.)

\	Damage Report			
Date	TI-15 Number	Class Period	Damage	Reported By

Introduction (cont.)

Utilizing and Managing the TI-15s in the Classroom (cont.)

Facilitating a TI-15 Center

If only a few calculators are available, create a TI-15 center in the back of the classroom. One suggested classroom layout is shown below. To prevent students from being distracted and to allow the teacher to work with small groups in the center while monitoring the other groups, have students sit in the TI-15 center with their backs to the other groups. Use the TI-15 Center Rotation Schedule below to keep track of which students have been to the center. While working with the students in the center, provide the other students with independent work. Use the *Check-Off List* (page 33) to keep track of which students used the calculators.

Classroom Layouts

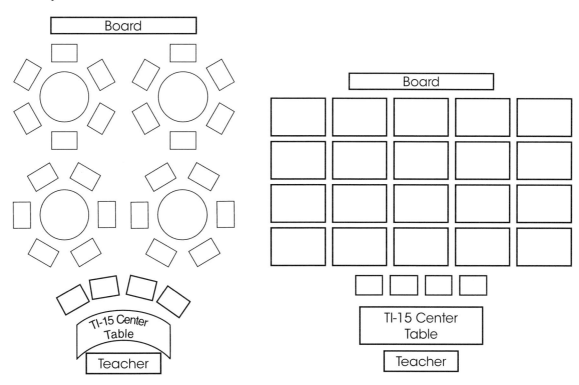

TI-15 Center Rotation Schedule

Days of the Week	Monday	Tuesday	Wednesday	Thursday	Friday
Group Names					
Students					

© Shell Education #50616—30 Mathematics Lessons Using the TI-15

Introduction (cont.)

Utilizing and Managing the TI-15s in the Classroom (cont.)

Parent Letter

A template of this letter is available on the Teacher Resource CD in both English and Spanish (letter.doc).

Greentree School
19685 Main Valley Road
St. Joseph, CA 92629

January 12, 2011

Dear Parents:

We have some exciting news to share with you about a new technology program we will be using in our classroom. Our class will be creating projects with the TI-15 Calculator, using the *30 Mathematics Lessons Using the TI-15*. We can select from many innovative and enjoyable activities. These projects combine subject-area learning with technology skill practice. With their clear, step-by-step directions, these projects are sure to be fun. We look forward to sharing our finished projects with you.

As calculators are becoming an important tool in mathematics at the elementary school level, this may be a great time to purchase a TI-15 calculator for your child. This would enable your child to practice at home the skills he or she has learned from the activities in school.

Sincerely,

_____

Miss Hood
3rd Grade Teacher

Introduction (cont.)

Assessment

For each lesson, activity sheets are provided that can be used to assess the students' knowledge of the concept. These activities could be considered as practice, in which students' progress and understanding of the concept is monitored through in-class assignments or homework. The activity sheets used in the Applying the Concept section of each lesson could also be assigned as a formal assessment.

Completion Grades

To give a completion grade for an activity, have students exchange papers. Review the problems together. Model problems on the overhead or have students model the problems. Have students count the number of problems completed. Then use the *Completion Grades Template* (page 38) to record students' scores. This template is also available on the Teacher Resource CD (grades.doc).

Write students' names above each column on the *Completion Grades Template*. Write the assignment title in the first column. Record students' scores as a fraction of the number of problems completed over the number of problems assigned. At the end of the grading period, add the number of problems completed for each student to the number of problems assigned. Divide the fraction to calculate a numerical grade.

Using a Point System for Formal Grades

When grading activities that serve as assessments, it is best to grade them yourself. This provides you with an opportunity to analyze students' performances, evaluate students' errors, and reflect on how instruction may have influenced their performances. It also prevents student error in grading. Depending on a school's grading procedures, assessments can be graded with fractions similar to completion grades. Determine the number of points that each problem is worth. You may want to assign two or more points for each problem if students are expected to show work or explain how they solved a problem. One point is awarded for the correct answer; the other points are for the students' work and/or written explanations. Write a fraction of the number of points a student earned out of the total number of points possible. Record these grades as fractions or convert them to percentages. Then, enter them into your grade book or an online grading system.

Grading with a Rubric

A rubric is an alternative way to grade activities or problems that involve multiple steps or tasks. It allows both the student and teacher to analyze a student's performance for the objectives of the task or assignment by giving the student a categorical score for each component. For example, if the students had the task of solving a problem and explaining how they solved it, a rubric would allow the teacher to identify in which subtasks students excelled or could improve. The problem may be correct, but the explanation may be missing certain steps needed to solve the problem. The *General Rubric* (page 39) and the *Create Your Own Rubric* (page 40) can be adapted for various types of activities. By using an all-purpose rubric, students can also be individually assessed on specific skills and objectives. Both of these rubrics are available on the Teacher Resource CD (genrubric.doc; rubric.doc).

Introduction (cont.)

Assessment (cont.)

Completion Grades Template

Student Names

Practice Activities

Total Points Earned										
Total Points Possible										

Key

Record Fractions for:
Number of Problems Completed
Number of Possible Problems

Introduction (cont.)

Assessment (cont.)

Directions: This rubric includes general criteria for grading multistep assignments that involve written explanations to questions. In each of the Level columns, specify each criterion by explaining how it relates to the activity and the levels of performance that can be achieved. Give the rubric to students for self-evaluation and peer evaluation. To evaluate an activity, circle a level of performance for each criterion and assign a number of points. Total the points and record them in one of the last three columns.

General Rubric

Criteria	Level I (0–4 points)	Level II (5–8 points)	Level III (9–10 points)	Self-Score	Peer Score	Teacher Score
Steps in the activity have been completed.						
Question(s) have been answered.						
Calculations are shown and/or explained.						
Responses relate to the questions being asked.						
Ideas are supported with logical reasoning and/or evidence.						

Introduction (cont.)

Assessment (cont.)

Directions: Write the criteria for the assignment in the first column. Then for each criterion, fill in the level of performance students may achieve. Give the rubric to the students for self-evaluation and peer evaluation. To evaluate an activity, circle a level of achievement for each criterion and then assign a number of points. Total the points and record them in one of the last three columns.

Create Your Own Rubric

Criteria	Level I (0–4 points)	Level II (5–8 points)	Level III (9–10 points)	Self-Score	Peer Score	Teacher Score

References Cited

Baker, S. K., D. C. Simmons, and E. J. Kame'enui. 1995. *Vocabulary acquisition: synthesis of the research.* (Tech. Rep. No. 13) Eugene: University of Oregon, National Center to Improve the Tools of Educators.

Beck, I. L., M.G. McKeown, and L. Kucan. 2002. *Bringing words to life: Robust vocabulary instruction.* New York: Guilford Press.

Burke, J. (No Date). Academic Vocabulary List. Obtained March 30, 2009 from http://www.englishcompanion.com.

Chval, K. B., and S. J. Hicks. 2009. Calculators in K–5 textbooks. *Teaching Children Mathematics* 15 (7): 430–37.

Echevarria, J., M. Vogt, and D. Short. 2004. *Making content comprehensible for English language learners: The SIOP model.* Boston: Pearson Education, Inc.

Ellington, A. 2003. A meta-analysis of the effects of calculators on students' achievement and attitude levels in precollege mathematics classes. *Journal for Research in Mathematics Education* 34 (5): 433–63.

Feldman, K., and K. Kinsella. 2005. *Narrowing the language gap: The case for explicit vocabulary instruction.* New York: Scholastic, Inc.

Frayer, D. A., W. D. Frederick, and H. J. Klausmeier. 1969. *A schema for testing the level of concept mastery.* Working Paper No. 16. Madison: Wisconsin Research and Development Center for Cognitive Learning.

Herrell, A., and M. Jordan. 2004. *Fifty strategies for teaching English language learners.* 2nd ed. Upper Saddle, NJ: Pearson Education, Inc.

Kutzler, B. 2000. The algebraic calculator as a pedagogical tool for teaching mathematics. *International Journal of Computer Algebra in Mathematics Education* 7 (1): 5–23.

National Council of Teachers of Mathematics. 2000. *Principles and standards for school mathematics.* Reston, VA: National Council of Teachers of Mathematics.

Reys, B., and F. Arbaugh. 2001. Clearing up the confusion over calculator use in grades K–5. *Teaching Children Mathematics* 8 (2): 90–94.

Ruthven, K., and D. Chaplin. 1997. The calculator as a cognitive tool: Upper-primary pupils tackling a realistic number problem. *International Journal of Computers for Mathematical Learning* 2: 93–124.

Sparrow, L., and P. Swan. 2005. Decimals, denominators, demons, calculators and connections. *Australian Primary Mathematics Classroom* 10 (3): 21–26.

Stiff, L. V. NCTM News Bulletin. April 2001.

Thompson, D. R., and R. N. Rubenstein. 2007. Learning mathematics vocabulary: Potential pitfalls and instructional strategies. *Mathematics Teacher* 93 (7): 568–74.

Lesson 1

Close Enough

Overview

Students learn to round numbers as they estimate the profits from a magazine sale.

Mathematics Objective

Uses specific strategies to estimate computations and to check the reasonableness of computational results.

TI-15 Functions

- Fix feature
- Operations

Materials

- TI-15s
- newspaper advertisements
- *Round It Up* activity sheet (pages 45–46; page045.pdf)
- *More Money* activity sheet (page 47; page047.pdf)

 **Vocabulary**

Complete the *Chart and Match* (page 19) vocabulary activity using the words below. Definitions for these words are included on the Teacher Resource CD (glossary.pdf).

- decimal point
- digit
- place value
- rounding

 Warm-up Activity

1. Ask students to enter a four-digit number beginning with 7 on their TI-15s.

2. Ask them to look at the value of the digit in the hundreds place. Have them raise their right hands if the value is less than five, raise their left hands if it is more than five, and stand if it equals five.

3. Have students press [Fix], [1000.], and [Enter] to round the number to the thousand's place. Ask students whose display shows 1000 to sit or put down their hands. Ask the class which of the original groups remain and what is on their displays.

4. Discuss how the rules for rounding explain how the [Fix] key changed the display. Have the students press [Fix] and [.] to return the TI-15s to the regular display.

5. Remind students that they need to press [On] and [Clear] at the same time to reset the memory of the TI-15.

Close Enough (cont.)

Explaining the Concept

1 Ask students to discuss the position of the digit that determines how a number is rounded. Have students investigate using prices in newspaper advertisements. (*The digit that is used to determine the rounding is the digit to the right of the place being rounded to.*)

2 Ask students to enter $23.42 onto their TI-15s. Remind them that the TI-15s will not display a dollar sign. Show students how to use the Fix feature to express the amount to the nearest dollar ([Fix] [.1] [Enter]). Explain that the TI-15 is rounding the number to the nearest dollar. Have students explain what digit determined whether the nearest dollar would be 23 or 24. (*The four in $23.42.*) Ask them the place value of the determining digit. (*tenths*) Have them try $23.64. Ask them what digit determines the answer. (*the six*) Teach them the rule for rounding to the nearest dollar: leave the number of dollars the same if the tenths digit is less than five. Increase it by one if the tenths digit is 5 or more.

3 Ask students: *Why might it be considered impossible to express $23.50 to the nearest dollar?* (*It is exactly halfway between.*) Have the students enter $23.50 into their TI-15s and press [Fix] [.1] [Enter]. Ask them what their TI-15s show. (*24*) Have them try other amounts where the number of cents is 50. Ask them how the TI-15 handles numbers that are exactly in the middle when rounding. (*It always rounds up.*) Explain that this is a rule we follow in mathematics.

4 Distribute copies of the *Round It Up* activity sheet (pages 45–46; page045.pdf). Ask students how to change the number of points to dollars and cents. Have them fill in the first two blanks on *Week One*, and then check their answers. Have them complete the rest of the amounts for the table.

5 To prepare students for the next part of the activity sheet, ask them whether they should use [Fix] and [.1] before or after they compute the values for the first method. To check, have part of the class use [Fix] and [.1] to calculate the value of $2.49 and $2.47. They will get $2 for each amount. Then have them add the two numbers together to get $2 + $2 = $4. Have the rest of the class compute the sum of $2.49 and $2.47 and then use [Fix] and [.1] on the sum. They will get $4.96 as the sum which the TI-15 will round to $5. Discuss why the answers are different.

6 Help students complete the information in Steps 2 and 3. Let them independently complete the rest of the two tables.

7 Allow time for students to answer questions 1–4 independently or in pairs. Discuss the answers as a class.

© Shell Education

Lesson 1

Close Enough (cont.)

Applying the Concept

1. Have students make and test a generalization of how the Fix feature will work with the [1000.], [100.], [10.], and [0.01] keys. Ask students: *If [Fix] [1.] rounds to the nearest ones, what will [Fix] [1000.], [100.], [10.], and [0.01] round to?*

2. Ask students: *How would we round a number if there were only dimes left in the world?* (to the nearest $0.10) Ask: *What key would we use with [Fix] to round to the nearest tenth?* ([0.1]) Remind them that [0.1] stands for one-tenth and that a dime is one-tenth of a dollar. Ask: *What place value and which digit do we look at to round to the nearest dime?* (*The hundredths place, which is the second digit to the right of the decimal.*)

3. Distribute copies of the *More Money* activity sheet (page 47; page047.pdf). Have students complete the activity sheet.

Differentiation

- **Below Grade Level—** Have students work in pairs to complete the *More Money* activity sheet. Use a number line to illustrate that rounding is really choosing the number that is closer. Walk students through the steps on the activity sheet.

- **Above Grade Level—** Ask students to repeat the activity from the *More Money* activity sheet, but this time tell students they received only 0.4 of a cent for each point. Have them invent a game that they can score, earn points, and calculate earnings. If time permits, have the pairs switch and solve each other's problems.

Extending the Concept

Have students investigate the use of tenths of a cent in the price of a gallon of gasoline. Have them determine which method of rounding is used to calculate the total cost of a tank of gasoline.

44 #50616—30 Mathematics Lessons Using the TI-15 © Shell Education

Name _____

Round It Up

Directions: Follow the steps below.

Step 1 Mrs. Romero's class is having a three-week magazine sale to earn money for a class trip to the zoo. Students earn points with every magazine subscription they sell. Each student receives one cent for each point. The points are listed below. Write the amount of money earned by each student next to the number of points in the table below.

Week One			Week Two			Week Three		
Student	Points	$ Total	Student	Points	$ Total	Student	Points	$ Total
Raj	372		Raj	1,055		Raj	619	
Jada	882		Jada	242		Jada	203	
Estelle	173		Estelle	147		Estelle	1,264	
Alex	951		Alex	328		Alex	746	

Step 2 The students decide to round the total amounts made over the three weeks to the nearest dollar. Jada wants to add up the amount for each week of the sale, and then round the answer. Use your TI-15 to find each student's total. First, to calculate the total before rounding, turn off the Fix feature by pressing [Fix] and [·]. Then after you have the total, press [Fix] [1.] and [Enter] to round it to the nearest dollar. Each time you begin calculations for another student, you will have to press [Fix] and [·] to turn off the Fix feature. Write your answers in the table below.

Student	$ Total	Method 1 (Rounded)
Raj		
Jada		
Estelle		
Alex		

© Shell Education   #50616—30 Mathematics Lessons Using the TI-15

Round It Up (cont.)

Directions: Follow the step below.

Step 3 Alex wants to use a different method of rounding. He wants to round the amount of each week per student first and then add the rounded amounts together to find the total. Use the [Fix] and [I.] keys on your TI-15 to round the totals for each week. Then add the rounded totals to find the approximate amount each student earned. Write your answers in the table below.

Student	Week 1	Week 2	Week 3	Method 2 (Rounded)
Raj				
Jada				
Estelle				
Alex				

Directions: Answer the questions below.

1. Were the totals always the same for the two methods?

2. Which student had different totals for the two methods? Why?

3. Why might you choose the first method?

4. Why might you choose the second method?

Name _____

More Money $

Directions: Follow the steps below.

Step 1 Mrs. Romero's class decided to extend the magazine sale for another three weeks. But this time, the students will round their earnings to the nearest dime, or $0.10. Their new points are shown below. Write the amount of money earned by each student next to the number of points in the table below.

Week One			Week Two			Week Three		
Student	Points	$ Total	Student	Points	$ Total	Student	Points	$ Total
Raj	859		Raj	2,232		Raj	674	
Jada	433		Jada	367		Jada	1,482	
Estelle	256		Estelle	319		Estelle	1,772	
Alex	418		Alex	955		Alex	289	

Step 2 Use your TI-15 to find each student's total. Then round to the nearest dime. Use the [Fix] and [0.1] keys on your TI-15. Write your answers in the table below.

Student	$ Total	Method 1 (Rounded)
Raj		
Jada		
Estelle		
Alex		

Step 3 Use your TI-15 to round each amount to the nearest dime for each week. Then add the rounded amounts to find the total. Write your answers below.

Student	Week 1	Week 2	Week 3	Method 2 (Rounded)
Raj				
Jada				
Estelle				
Alex				

Which students had different totals for method 1 and method 2?

_____

Lesson 2

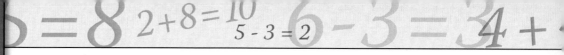

Leftovers

Overview

Students will learn about the meaning of the remainder in a division problem by dividing candy between students and a teacher.

Mathematics Objective

Understands the concept of division. Multiplies and divides whole numbers.

TI-15 Functions

- Fix feature
- Integer division
- Operations

Materials

- TI-15s
- *Divvy It Up!* activity sheet (pages 51–52; page051.pdf)
- *By the Numbers* activity sheet (page 53; page053.pdf)

Vocabulary

Complete the *Gouin Series* (page 18) vocabulary activity using the words below. Definitions for these words are included on the Teacher Resource CD (glossary.pdf).

- dividend
- divisor
- per
- quotient
- remainder

Warm-up Activity

1. Ask each student to write down a division problem with a divisor of 5. The answer should not have a remainder. Ask them to check their problems on their TI-15s using the ÷ key. Have them repeat the computation using the Int÷ key. Discuss the meaning of the r ▯ on the display.

2. Have students make a generalization about the difference in functionality between the ÷ and Int÷ keys. Explain that when the Int÷ key is used, the answer is displayed with a quotient and a remainder.

3. Have students add 1 to the dividend of their division problems and use the Int÷ key again.

4. Ask one student to give the remainder. Have all students with the same remainder raise their hands.

5. Have them add 2 to their original dividend and use the Int÷ key again and check the remainders.

6. Continue with this activity by having students add 3, 4, 5, and 6 to their original dividends.

7. Remind students that they need to press On and Clear at the same time to reset the memory of the TI-15.

48   #50616—30 Mathematics Lessons Using the TI-15   © Shell Education

Leftovers (cont.)

Explaining the Concept

1 Ask the class: *Have you ever tried to evenly divide a bag of candy among a group of friends? How did you decide how many pieces each friend received? What did you do with the leftover pieces?* Explain to students that they are going to investigate dividing bags of candy among a group of students and find out how many pieces are left over.

2 On the board or overhead, draw the chart below. Tell students they need to fairly divide 25 pieces of candy between 2 students. Have them draw a picture to show the number of pieces per student and the number left over. Have them identify the divisor, dividend, quotient, and remainder. So, with 25 pieces (dividend) shared between 2 students (divisor), each student receives 12 pieces (quotient) with 1 piece left over (remainder).

Pieces in Bag	# of Students	Pieces per Student	Pieces Left Over

3 Ask students to change the number of pieces to 26 and have them divide the candy among 2, 3, and then 4 students. *(13 r 0, 8 r 2, 6 r 2)* Have students record their results. Ask students: *How can you use the TI-15 to find the number of pieces per student and the number left over?* (Use the [Int÷] key.) Have them use their TI-15s to check their results on the chart.

4 Distribute copies of *Divvy It Up* (pages 51–52; page051.pdf). Explain that the teacher, Mr. Kuo, will receive any leftover pieces. Discuss the division problem needed to find the number of pieces of candy for each student and how much is left for Mr. Kuo.

5 After students have completed the chart on *Divvy It Up*, illustrate the meaning of the answers given on the TI-15s by choosing one of the problems with a remainder and drawing the 8 groups with the appropriate amount of candy in each as well as by drawing the remainder of the candy that goes to Mr. Kuo.

6 Discuss the dividend (total number of pieces of candy), the divisor (the 8 students), the quotient (number of pieces of candy for each student), and the remainder (the number of pieces of candy left over) using the drawing. Emphasize that you are dividing by 8 because you are finding the number of pieces of candy per student, and there are 8 students. Have students complete the *Divvy It Up!* activity sheet.

7 Review the answers for page 52. Ask students: *Why will Mr. Kuo never receive nine pieces?* Use the diagram to demonstrate that if there were nine pieces left over, another piece of candy could have been given to each student.

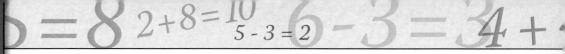

Lesson 2

Leftovers (cont.)

Applying the Concept

1. Distribute copies of the *By the Numbers* activity sheet (page 53; page053.pdf), and have students complete them. Remind students how to press [Fix] [I.] to find the answer to a problem to the nearest whole number. Show them that while the [÷] key is affected by the [Fix] [I.] setting, the [Int÷] key is *not* affected by it.

2. Discuss how to compare the size of the remainder with half the divisor to tell if a quotient will be rounded up or down.

3. Remind students if they press the wrong key, they can use the [←] to go back to the place of the error and correct it. Or they can press (Clear) to start over again.

Differentiation

- **Below Grade Level—** Have students work in pairs, and use blocks or pieces of candy to form groups and find remainders for the problems on the *By the Numbers* activity sheet.

- **Above Grade Level—** Have students look at the decimal approximations of the quotients on the *By the Numbers* activity sheet and find the decimal values for several divisors and remainders.

Extending the Concept

Create fractions out of the division problem's solution and use the remainder as the numerator and the divisor as the denominator. Determine when the value of the fraction is less than 0.5 and when it is greater than or equal to 0.5.

Name _____

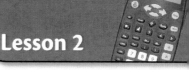

Divvy It Up!

Directions: Follow the steps below and then answer the questions on the next page.

The number of pieces of candy in a bag may vary from 49 to 64. Eliza is going to divide the candy evenly among the 8 students in the classroom. The teacher, Mr. Kuo, will get whatever is left. Use the [Int÷] key on your TI-15 to help you fill out the chart below.

Number of Pieces in Bag	Pieces per Student	Pieces for Mr. Kuo
49		
50		
51		
52		
53		
54		
55		
56		
57		
58		
59		
60		
61		
62		
63		
64		

© Shell Education

#50616—30 Mathematics Lessons Using the TI-15

Divvy It Up! (cont.)

Directions: Use the chart you just completed to answer the questions below.

1. What is the divisor of the division problems? Why is it the divisor? _____

   _____

2. Which column shows the dividend of the division problem?

   _____

3. Which column shows the quotient of the division problem? _____

4. Which column shows the remainder of the division problem? _____

5. What is the largest number of pieces Mr. Kuo could receive? _____

6. What is the smallest number of pieces Mr. Kuo could receive? _____

7. Did Mr. Kuo ever get more candy than the other students? _____

8. Did Mr. Kuo ever get less candy than the other students? _____

9. Describe the different ways that the remaining number of pieces for Mr. Kuo could change if there were more pieces of candy in the bag. You may find it helpful to give examples.

   _____

   _____

   _____

   _____

10. In the chart on the previous page, what are all the possible number of pieces of candy Mr. Kuo might receive?

    _____

Name _____

By the Numbers

Directions: Help Mr. Kuo predict how many pieces of candy he may receive by answering the questions below.

1. Suppose there are 105 to 119 pieces of candy in a bag, and there are 15 students.

    a. How many pieces would each student get? _____

    b. What are the possible numbers of pieces Mr. Kuo could get? _____

    c. Could the number of pieces left for Mr. Kuo be greater than the number each of the other students get?

    _____

    d. Which dividend will give the same number of pieces of candy to the students and to Mr. Kuo?

    _____

2. Suppose there are 24 to 26 pieces in a bag, and there are 3 students.

    a. How many pieces would each student get? _____

    b. What are the possible numbers of pieces Mr. Kuo could get? _____

    c. Could the number of pieces left for Mr. Kuo be greater than the number the other students get?

    _____

One day, Mr. Kuo knew that there were 12 students in the classroom, and the number of pieces of candy in the bag was 64 to 68.

3. What do you know about the number of pieces each student would receive?

    _____

4. What do you know about the number of pieces that would be left for Mr. Kuo?

    _____

5. Use the [Fix] [l.] and [÷] keys to divide each possible amount of candy by 12. How many pieces of candy could Mr. Kuo receive when the rounded quotient is 5? What about when the rounded quotient is 6?

    _____

Lesson 3

Half Again

Overview

Students will learn how to find fractions whose sum is one-half using fractions with even and odd denominators. They will investigate common denominators and simplify fractions.

Mathematics Objective

Adds and subtracts simple fractions. Understands the concept of addition of fractions. Understands the equivalence of fractions.

TI-15 Functions

- Entering fractions
- Operations
- Simplifying fractions

Materials

- TI-15s
- *It All Adds Up* activity sheet (pages 57–58; page057.pdf)
- *Different Halves* activity sheet (page 59; page059.pdf)

Vocabulary

Complete the *Sentence Frames* (page 16) vocabulary activity using the words below. Definitions for these words are included on the Teacher Resource CD (glossary.pdf).

- denominator
- fraction
- numerator
- equivalent
- lowest terms
- simplify

Warm-up Activity

1. On the board or overhead, draw a number line showing the interval between zero and one. Label the zero and one.

2. Divide the unit into eighths. Ask students how many sections the number line is divided into. *(eight)*

3. To display the distance covered on the number line, solve the equation $0 + \frac{1}{8}$. Have students press [0] [+] [1] [n] [8] [d] [Enter] to display the distance covered from 0 to the first mark on the number line. Write $\frac{1}{8}$ on the number line where appropriate.

4. Ask students what they need to add to $\frac{1}{8}$ to get to the next section on the number line. $(\frac{1}{8})$

5. Have students press [+] [1] [n] [8] [d] [Enter] to add $\frac{1}{8}$ to the previous number. Then ask if that fraction is in lowest terms. *(no)* Students should press [Simp] and then [Enter] to simplify the number. $(\frac{1}{4})$

6. Write the fraction in lowest terms on the number line in the appropriate place.

7. Repeat adding eighths, simplifying, and writing on the number line until the number line is complete.

54 #50616—30 Mathematics Lessons Using the TI-15 © Shell Education

Half Again (cont.)

Explaining the Concept

1 Ask the class: *Suppose a pizza was divided into four pieces and there was half of it left. How many pieces are left?* On a sheet of paper, have students draw the whole pizza and divide it into fourths. Then have them shade half of the pizza. Discuss how $\frac{1}{4} + \frac{1}{4} = \frac{2}{4} = \frac{1}{2}$. Use the sketch to clarify why $\frac{1}{4}$ represents one out of the four equal pieces that make up the whole pizza. Explain that $\frac{2}{4}$ represents the shaded region, and how $\frac{2}{4}$ and $\frac{1}{2}$ are equivalent fractions. Repeat with a pizza divided into 8 pieces, showing students that $\frac{2}{8} + \frac{2}{8} = \frac{4}{8} = \frac{1}{2}$.

2 On the board or overhead, draw three pieces of pizza that when combined, equal $\frac{1}{4}$ of a pizza. Identify each piece as $\frac{1}{12}$ of the pizza, but do not give the combined value of $\frac{1}{4}$. Ask students: *How many pieces are needed to complete half of the pizza?* (3 pieces) Ask: *How do we represent the 3 pieces as a fraction of the pizza?* ($\frac{3}{12}$) Help students find the sum of $\frac{3}{12} + \frac{3}{12}$. ($\frac{6}{12}$) Then have students represent the fraction $\frac{6}{12}$ in multiple ways. (6 out of 12 pieces, $\frac{1}{2}, \frac{3}{6}, \frac{2}{4}$) Discuss how it was known that if 12 pieces make up the whole pizza, a total of 6 pieces would be needed for $\frac{1}{2}$ of the pizza.

3 Explain to students that the class will be having a contest to see which half of the class can find the greatest number of correct addends to equal $\frac{1}{2}$. Distribute copies of *It All Adds Up* (pages 57–58; page057.pdf) to students. Tell them that they are going to find the missing addends for the first part of the activity sheet without using their TI-15s, and then exchange papers with partners. Each partner will use the TI-15 to check the answers.

4 Give students a pre-determined amount of time to complete the first set of problems. After they have exchanged papers, show them how to enter $\frac{1}{4}$ by pressing [1] and then [n] to enter the 1 as the numerator of a fraction, and then press [4] to enter the denominator. When they press [+], the 4 is set as the denominator. Enter $\frac{1}{4}$ again and press [Enter]. The TI-15 displays $\frac{2}{4}$. Press [Simp] and [Enter] to simplify the fraction to $\frac{1}{2}$. Have students check the first set of problems. Then have them return the activity sheets to their partners.

5 Give students a predetermined amount of time to complete the second part of the activity sheet. Have students exchange activity sheets again and use their TI-15s to check their partner's work. After they have written the total number of points on the paper, have students return the activity sheets to their partners. Allow students time to complete the remaining questions. Discuss the answers to the questions on the activity sheet as a class.

6 Have students think about and generalize the mathematics they used by answering the questions in the last part of the activity sheet.

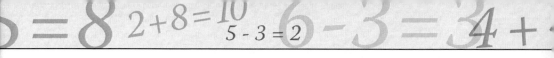

Lesson 3

Half Again (cont.)

Applying the Concept

1. Show students an interval on a number line divided into 3 parts. Label one part $\frac{1}{3}$ and the other part $\frac{2}{3}$. Ask students: *Where does $\frac{1}{2}$ belong on this line?* They should suggest halfway between the $\frac{1}{3}$ and the $\frac{2}{3}$ marks. Mark the $\frac{1}{2}$ and divide the other segments in half to form six parts. Ask students how many pieces the interval is divided into now.

2. Ask students what fraction represents the first mark after zero. ($\frac{1}{6}$) Ask them how far they need to move to get to $\frac{1}{2}$. Have students write $\frac{1}{6} + \frac{2}{6}$ and have them identify the sum in two different ways. ($\frac{3}{6}$ and $\frac{1}{2}$) Have students enter $\frac{2}{6}$ into the TI-15 and press [Simp] and [Enter]. Ask them to write a number sentence using $\frac{1}{3}$. ($\frac{1}{3} + \frac{1}{6} = \frac{1}{2}$) Distribute copies of *Different Halves* (page 59; page059.pdf). Tell students to follow the calculator steps slowly and carefully. Have them raise their hands if they need help using their TI-15s.

Differentiation

- **Below Grade Level—** Have students make extensive use of sketches while working on the *Different Halves* activity sheet. Adjust the assignment so that students not only consider $\frac{1}{5} \times \frac{2}{2}$ in question 2, but that they also consider $\frac{1}{5} \times \frac{3}{3}$, $\frac{1}{5} \times \frac{4}{4}$, and $\frac{1}{5} \times \frac{20}{20}$. You may wish to have students complete *Different Halves* with a partner.

- **Above Grade Level—** Have students explore the [Simp] key further. For example, when simplifying $\frac{5}{25}$, they can enter the fraction and then press [Simp] [5] and [Enter] to immediately find the simplified fraction $\frac{1}{5}$. Have them create problems in which two addends give a sum of $\frac{3}{4}$. Encourage the use of two-digit denominators.

Extending the Concept

Have students look at the connection between fractions and percents. For example, if 24% of a group of students are wearing jeans and there are 100 students, then 24 of them are wearing jeans. If there are 50 students, then 12 of them have jeans on because 24/100 = $\frac{12}{50}$. Ask: *How many students have jeans on if there are 25 students?* Have them use the [Simp] key to help with that part of the problem. Instruct students to find or think of examples where percents can be changed to fractions and then those fractions can be simplified. They should be able to explain those examples just as they can explain the 25% example.

Name _____

Lesson 3

It All Adds Up

Directions: Do not use your TI-15 for this part of the activity. For each fraction in the chart below, find the missing fraction that would make the sum of the two fractions equal to $\frac{1}{2}$. Complete as many as possible before your teacher calls "Time's up." Then, follow the directions below.

1. $\frac{1}{4}$ + ☐ = $\frac{1}{2}$	2. $\frac{2}{8}$ + ☐ = $\frac{1}{2}$	3. $\frac{3}{12}$ + ☐ = $\frac{1}{2}$
4. $\frac{4}{16}$ + ☐ = $\frac{1}{2}$	5. $\frac{5}{20}$ + ☐ = $\frac{1}{2}$	6. $\frac{6}{24}$ + ☐ = $\frac{1}{2}$
7. $\frac{7}{28}$ + ☐ = $\frac{1}{2}$	8. $\frac{8}{32}$ + ☐ = $\frac{1}{2}$	9. $\frac{7}{32}$ + ☐ = $\frac{1}{2}$
10. $\frac{10}{40}$ + ☐ = $\frac{1}{2}$	11. $\frac{20}{80}$ + ☐ = $\frac{1}{2}$	12. $\frac{21}{80}$ + ☐ = $\frac{1}{2}$

Directions: Now trade papers with a partner. Use your TI-15 and follow the steps below to check your partner's answers. Score 2 points for each correct fraction.

Step 1 Enter the numerator of the first fraction. Press [n].

Step 2 Enter the denominator of the first fraction. Press [+].

Step 3 Enter the numerator of the second fraction. Press [n].

Step 4 Enter the denominator of the second fraction. Press [Enter].

Step 5 Press [Simp]. Press [Enter]. You many need to press [Simp] and [Enter] more than once to arrive at the lowest terms.

Total points: _____

Now return the paper to your partner.

© Shell Education #50616—30 Mathematics Lessons Using the TI-15

It All Adds Up (cont.)

Directions: Do not use your TI-15 for this part of the activity. For each fraction in the chart below, find the missing fraction that would make the sum of the two fractions equal to $\frac{1}{2}$. Complete as many as possible before your teacher calls "Time's up." Then, follow the directions below.

1. $\frac{2}{6} + \square = \frac{1}{2}$	2. $\frac{3}{10} + \square = \frac{1}{2}$	3. $\frac{3}{6} + \square = \frac{1}{2}$
4. $\frac{4}{12} + \square = \frac{1}{2}$	5. $\frac{1}{6} + \square = \frac{1}{2}$	6. $\frac{3}{14} + \square = \frac{1}{2}$
7. $\frac{5}{18} + \square = \frac{1}{2}$	8. $\frac{7}{20} + \square = \frac{1}{2}$	9. $\frac{4}{24} + \square = \frac{1}{2}$
10. $\frac{1}{16} + \square = \frac{1}{2}$	11. $\frac{24}{50} + \square = \frac{1}{2}$	12. $\frac{99}{200} + \square = \frac{1}{2}$

Directions: Now trade papers with a partner again. Use your TI-15 and follow the steps below to check your partner's answers. Score 2 points for each correct fraction.

- **Step 1** Enter the numerator of the first fraction. Press [n].
- **Step 2** Enter the denominator of the first fraction. Press [+].
- **Step 3** Enter the numerator of the second fraction. Press [n].
- **Step 4** Enter the denominator of the second fraction. Press [Enter].
- **Step 5** Press [Simp]. Press [Enter]. You many need to press [Simp] and [Enter] more than once to arrive at the lowest terms.

Total points: _____

Directions: Return the paper to your partner and then answer the questions below.

1. What was always true about the numerator and denominator of the fraction before you pressed [Simp]? _____

2. What did pressing [Simp] do to the fraction? _____

3. Explain how you knew what fraction to add. _____

Name _____

Different Halves

Lesson 3

Directions: Follow the steps below.

1. Enter the fraction $\frac{2}{2}$ into your TI-15. Press [Enter]. What does the display show?
   _____

2. Use your TI-15 to solve this problem without simplifying: $\frac{1}{5} \times \frac{2}{2}$. What does the display show?
   _____

3. How is this fraction related to $\frac{1}{5}$? _____

4. Why is this true? _____
   _____

5. What fraction could you add to $\frac{1}{5}$ so that the sum is equal to $\frac{1}{2}$? _____

6. Check the sum on your TI-15. Was your answer correct? _____

7. Use the TI-15 to find a fraction equal to $\frac{3}{7}$ with a denominator of 14. _____

8. What fraction could you add to $\frac{3}{7}$ so that the sum is equal to $\frac{1}{2}$? _____

9. Check your answer on your TI-15. Was your answer correct? _____

10. What fraction would you add to $\frac{4}{13}$ so that the sum is $\frac{1}{2}$? _____

    Explain how you found your answer.
    _____
    _____
    _____
    _____

Lesson 4

More or Less

Overview

Students will learn to compare the values of proper fractions.

Mathematics Objective

Understands the concept of proper fractions.

TI-15 Functions

- Entering fractions
- Inequalities
- Operations

Materials

- TI-15s
- *Fair Prices* activity sheet (pages 63–64; page063.pdf)
- *Who Has More?* activity sheet (page 65; page065.pdf)

Vocabulary

Complete the *Vocabulary Bingo* (page 19) vocabulary activity using the words below. Definitions for these words are included on the Teacher Resource CD (glossary.pdf).

- compare
- denominator
- equivalent
- fraction
- inequality
- least common denominator
- numerator

Warm-up Activity

1. Tell students: *Today we are going to compare fractions.* Display two fractions with common denominators, such as $\frac{3}{4}$ and $\frac{1}{4}$.

2. Ask students to predict which fraction has the larger value. Tell students they will use the division function and inequality keys to test fractions' inequalities on the TI-15.

3. To check their predictions, have students press ◆ and then (Mode). Then press ⇨ to move the cursor under MAN. Press [Enter] and then (Mode). This puts the TI-15 in manual problem-solving mode.

4. Ask students: *How can $\frac{3}{4}$ be shown using division?* Have them type this first fraction using division by pressing [3] [÷] [4].

5. Next have students press ⟨⟩ once or twice depending on whether they think the first fraction is smaller or larger.

6. Then students must type the second fraction using division again ([1] [÷] [4]).

7. Finally, have students press [Enter]. The TI-15 will then tell students if their predictions were correct.

8. Repeat this using other fractions with common denominators.

Lesson 4

More or Less (cont.)

Explaining the Concept

1 Ask the class: *Suppose there are two pies of equal size. One pie is cut into 6 pieces and the other pie is cut into 8 pieces. If you could have just one piece of pie, how would you choose which pie to take the piece from so that you had the most pie?*

2 Have students draw a picture of the scenario to predict which pie they would choose their piece from. Explain to students that they need to compare $\frac{1}{6}$ and $\frac{1}{8}$. Have them sketch pies to see that $\frac{1}{6} > \frac{1}{8}$. Explain that to mathematically check their prediction, they must find equivalent fractions with a common denominator. Have students list multiples of 6 and 8 and identify the lowest common denominator. (24) Determine which factors to multiply by to get the common denominator. ($\frac{1}{6} \times \frac{4}{4} = \frac{4}{24}$ and $\frac{1}{8} \times \frac{3}{3} = \frac{3}{24}$) Ask students if their predictions were correct. Try comparing other fractions with unlike denominators and with numerators greater than 1.

3 Ask students to consider if they had two pieces of the first pie and five pieces of the second pie. Looking at the picture should convince them that $\frac{2}{6}$ is less than $\frac{5}{8}$. Show them how to write the inequality $\frac{2}{6} < \frac{5}{8}$. Point out that the inequality $\frac{5}{8} > \frac{2}{6}$ gives the same information. Have students check the inequalities by hand as described in Step 2.

4 Show students how to test the inequalities on the TI-15. Press ◈ and then `Mode`. Press ▷ once to move the cursor under MAN. Press `Enter` and then `Mode`. This allows the student to enter a statement to be tested as true (Yes) or false (No). Then, press `5` `÷` `8` `<>` `<>` `2` `÷` `6` `Enter`. Explain that pressing the `<>` key once, indicates less than, and twice, indicates greater than. The manual problem-solving mode will not work with fraction notation, so fractions will be represented using division. Have students practice changing the order of the inequality. Have them enter $\frac{2}{6} < \frac{5}{8}$ by pressing `2` `÷` `6` `<>` `5` `÷` `8` `Enter`. The TI-15 will display `YES`. Have students press `On` and `Clear` at the same time to reset the memory of the TI-15.

5 Distribute *Fair Prices* (pages 63–64; page063.pdf). Explain to students they are going to determine which fair vendor will sell you the most pie at the lowest price. Help students work through the first section. When finding equivalent fractions in part 1c, show students how to enter $\frac{2}{5}$ by pressing `2` `n` `5`. Pressing `×` moves the cursor off of the denominator. Have them enter $\frac{7}{7}$ by pressing `7` `n` `7` `Enter`. The answer is $\frac{14}{35}$. Make sure they understand that multiplying by $\frac{7}{7}$ is the same as multiplying by 1 and gives an answer that is an equivalent fraction. Have students complete the rest of the activity sheet.

Lesson 4

More or Less (cont.)

Applying the Concept

1. Distribute copies of the *Who Has More?* activity sheet (page 65; page065.pdf), and have students complete the activity. Tell them that they will be finding a method for comparing fractions.

2. When they have finished the activity sheet, ask them to describe the general method for comparing the fractions. They should know that they are multiplying the numerator and denominator of each fraction by the denominator of the other.

Differentiation

- **Below Grade Level**—Give students paper copies of pies marked into the number of pieces needed for the common denominator. Have them model problems by cutting out pieces and matching them on the models.

- **Above Grade Level**—Have students find the least common multiple of the denominators and use the lowest common denominator to compare the fractions.

Extending the Concept

Ask students if it is necessary to find the common denominator to determine which fraction is larger. Show them the "cross product" method to compare the fractions, where they multiply the numerator of the first fraction with the denominator of the second fraction. For example, from the *Who Has More?* activity sheet, the fractions that describe Eli's pie and Jill's pie are $\frac{4}{13}$ and $\frac{9}{25}$ respectively. Look at the fractions written side by side. Multiply across the X, placing each product with the numerator used. Have students write a paragraph explaining why this method works.

$$100 \quad 117$$
$$\frac{4}{13} \times \frac{9}{25}$$

Name _____

Fair Prices

Lesson 4

Directions: Answer the questions below.

Your class takes a trip to the local fair. There is a booth selling pieces of pie and cake. Try to choose the deal that is going to get you the most pie for the lowest price.

1. **First Deal:** You can buy 2 pieces of apple pie from a pie that is cut into 5 pieces.
   Second Deal: You can buy 3 pieces of cherry pie from a pie that is cut into 7 pieces.

   a. What fraction of the apple pie would you get? _____

   b. What fraction of the cherry pie would you get? _____

   c. Use the TI-15 to multiply $\frac{2}{5}$ by $\frac{7}{7}$. What does the TI-15 show? _____

   d. How does the value of $\frac{2}{5}$ compare to the value of $\frac{14}{35}$? _____

   e. Multiply $\frac{3}{7}$ by $\frac{5}{5}$. What does the TI-15 show? _____

   f. How does the value of $\frac{3}{7}$ compare to the value of $\frac{15}{35}$? _____

   g. Compare the fractions. Which deal would get you more pie for your money?

   _____

   h. Write two inequalities comparing the original fractions.

   _____

Directions: Now use the TI-15 and the steps below to check your inequalities from problem h. If your inequalities are not correct, adjust your answers above.

   Step 1   Press ◈ (Mode).
   Step 2   Press ▷. This moves the line under MAN.
   Step 3   Press [Enter] (Mode).
   Step 4   Enter the numerator of the first fraction.
   Step 5   Press [÷].
   Step 6   Enter the denominator of the first fraction.
   Step 7   Press ◇ once for **less than**. Press ◇ twice for **greater than**.
   Step 8   Enter the numerator of the second fraction.
   Step 9   Press [÷].
   Step 10 Enter the denominator of the second fraction. Press [Enter].

© Shell Education

Lesson 4

Fair Prices (cont.)

Directions: Answer the questions below.

2. **First Deal:** You can buy 6 pieces of chocolate cake from a cake that is cut into 7 pieces.
   Second Deal: You can buy 7 pieces of lemon cake from a cake that is cut into 9 pieces.

   a. What fraction of the chocolate cake would you get? _____

   b. What fraction of the lemon cake would you get? _____

   c. Use the TI-15 to multiply $\frac{6}{7}$ by $\frac{9}{9}$. What does the TI-15 show? _____

   d. How does the value of $\frac{6}{7}$ compare to the value of $\frac{54}{63}$? _____

   e. Use the TI-15 to multiply $\frac{7}{9}$ by $\frac{7}{7}$. What does the TI-15 show? _____

   f. Why is $\frac{7}{7}$ chosen? _____

   g. Compare the fractions. Which deal would get you more cake?
   _____

   h. Write two inequalities comparing the original fractions.
   _____

Directions: Now use the TI-15 and the steps below to check your inequalities from problem h. If your inequalities were not correct, adjust your answers above.

Step 1	Press ◈ Mode.
Step 2	Press ▶. This moves the line under MAN.
Step 3	Press Enter Mode.
Step 4	Enter the numerator of the first fraction.
Step 5	Press ÷.
Step 6	Enter the denominator of the first fraction.
Step 7	Press ◁ once for **less than**. Press ▷ twice for **greater than**.
Step 8	Enter the numerator of the second fraction.
Step 9	Press ÷.
Step 10	Enter the denominator of the second fraction. Press Enter.

64 #50616—30 Mathematics Lessons Using the TI-15 © Shell Education

Name _____

Who Has More?

Lesson 4

Directions: Follow the steps to help you answer each question.

1. Elena has $\frac{5}{12}$ of a pie and Javier has $\frac{7}{18}$ of a pie.

    a. What fraction equivalent to 1 should you multiply Elena's fraction by to find a common denominator with Javier's fraction?
    _____

    b. Use the TI-15 to multiply $\frac{5}{12}$ with the fraction in part a. What does the TI-15 show?
    _____

    c. What fraction equivalent to 1 should you multiply Javier's fraction by to find a common denominator with Elena's fraction?
    _____

    d. Use the TI-15 to multiply $\frac{7}{18}$ with the fraction in part c. What does the TI-15 show?
    _____

    e. Who has more pie? _____

    f. Write two inequalities that compare $\frac{5}{12}$ and $\frac{7}{18}$. _____

    g. Now use your TI-15 to check your answer. Was your answer correct? _____

2. Eli has $\frac{4}{13}$ of a pie and Jill has $\frac{9}{25}$ of a pie.

    a. What fraction equivalent to 1 should you multiply Eli's fraction by to find a common denominator with Jill's fraction?
    _____

    b. What fraction equivalent to 1 should you multiply Jill's fraction by to find a common denominator with Eli's fraction?
    _____

    c. Which fraction is larger: $\frac{4}{13}$ or $\frac{9}{25}$? _____

    d. Who has more pie? _____

    e. Write two inequalities that compare $\frac{4}{13}$ and $\frac{9}{25}$. _____

    f. Now use your TI-15 to check your answer. Was your answer correct? _____

Lesson 5

Big Buck$

Overview

Students will learn to express products of repeated factors with exponents and evaluate exponential expressions.

Mathematics Objective

Understands the concept of exponents.

TI-15 Functions

- Constant feature
- Raising to a power

Materials

- TI-15s
- *Uncle Buck$* activity sheet (pages 69–70; page069.pdf)
- *Dollars and Days* activity sheet (page 71; page071.pdf)

Vocabulary

Complete the *Frayer Model* (page 20) vocabulary activity using the words below. Definitions for these words are included on the Teacher Resource CD (glossary.pdf).

- base
- exponent
- factor
- power

Warm-up Activity

1. Have students press [Opl] [+] [1] [Opl]. Write *0 + 1 = 1* on the board or overhead.

2. Ask them to press [0] [Opl].

3. Have students press [Opl] again. Write *1 + 1 = 2*. Ask students if they notice that the display shows how many times they have pressed [Opl], and that this is the same as the number of ones that have been added. Ask them how to express the answer as the product of this number and 1. (*2 x 1 = 2*)

4. Ask them to continue to press [Opl] using the display to keep track of the number of ones that have been added. Have them write each answer as a product of that number and 1 (e.g., 1 x 1 = 1, 2 x 1 = 2, 3 x 1 = 3, 4 x 1 = 4).

5. Have them press [On] and [Clear] at the same time to clear the constant feature. Have them repeat the exercise with twos instead of ones, and then threes instead of ones.

6. Ask the students how to express repeated addition as multiplication. (For example, 2 + 2 + 2 + 2 = 4 x 2.)

Big Buck$ *(cont.)*

Explaining the Concept

1 Ask the class: *What would happen if we kept multiplying by the same number instead of adding? Would the answer be the same as if we added?*

2 Pick 10 students and tell them that you are going to start with a number and then as you call on them, they must double the previous number. Start with the number 1. The first student should respond "2," the next should respond "4," the next "8," etc. The tenth student should respond with "1,024." Students may use their TI-15s during this exercise.

3 Draw the chart below on the board or overhead. Show students how to use constant feature on the TI-15 to complete repeated computations. Have them press [Op1] [×] [2] [Op1], and then press [1] [Op1]. Keep track of the results and the number of times [Op1] has been pressed after the 1 on the chart. Ask them what the number of times [Op1] has been pressed represents. *(It's the number of times 2 has been used as a factor.)*

# of times pressed	0	1	2	3	4	5	6	7	8	9
Result										

4 Explain to students that they are going to investigate what happens when the same number is used as a factor repeatedly and how to write an expression that represents it. Distribute copies of the *Uncle Buck$* activity sheet (pages 69–70; page069.pdf). Have students use their TI-15s as they fill in both charts.

5 When students are finished and the answers are checked, discuss the number of times that the repeated factor is used. Explain that there is a shortcut to write the expressions. Explain that the shortcut is the use of exponents. Show them how to write the expression 2 x 2 x 2 x 2 as 2^4. Explain that the 2, the repeated factor, is called the *base* and the 4, the number of times the factor is repeated, is called the *exponent*. Tell them that this is read as *2 to the fourth power.*

6 Show students how to write 3 x 3 x 3 x 3 x 3 x 3 as 3^6. Ask them to identify the base and the exponent. Ask them how to read the expression. Then ask students how to read the expression 5^3. Have them identify the base and the exponent and write the expression as repeated factors. Ask them to find the value of the expression. *(5 x 5 x 5 = 125)*

7 Show students the [^] key. Ask them to press [2] [^] [4] [Enter]. In parts a–e of the activity sheet, have students write each expression in exponential form and then simplify by using the [^] key. Point out that given only one factor of 2, the exponential representation for it is 2^1 and the simplified form is 2.

Lesson 5

Big Buck$ (cont.)

Applying the Concept

1 Distribute copies of the *Dollars and Days* activity sheet (page 71; page071.pdf), and have students complete them.

2 Remind students to clear the [Op1] by pressing [On] and [Clear] at the same time.

Differentiation

- **Below Grade Level**—Have students write the exponential products using a triangular display.

- **Above Grade Level**—Ask students to consider the meaning of 2^0 by looking at the date of birth as year 0. Show them that given a number like $3^3 = 27$, the next smaller value is found by dividing by the base: $3^2 = 27 \div 3$. Continue the pattern to look at 3^0, 3^{-1}, and 3^{-2}.

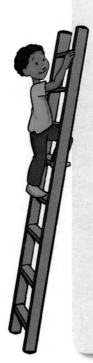

Extending the Concept

Have students investigate powers of 10 using [Op1] and [^]. Ask them to consider both the number of zeros and also the movement of the decimal point as the exponent decreases. Ask them to speculate on the meaning of 10^0 and 10^{-1}.

Name _____

Uncle Buck$

Directions: Read the text, fill in the chart, and answer the questions below.

1. Hannah's Uncle Buck$ gave her $1 on the day she was born. On each of her birthdays, he will give her twice as much as he did the year before until she is 26.

Use your TI-15 and the constant feature (Op1) to find how much she will be given on each birthday. On your TI-15, press [Op1] [×] [2] [Op1]. Press [1] and then [Op1]. Continue to press [Op1] to find the amount she will receive on each birthday. Write the amounts of the gifts on the chart.

Age	$$$	Age	$$$	Age	$$$
0		9		18	
1		10		19	
2		11		20	
3		12		21	
4		13		22	
5		14		23	
6		15		24	
7		16		25	
8		17		26	

Directions: Now, write the multiplication problem that would give the amount she receives on each of the birthdays below. Then give the number of times 2 is used as a factor in each product.

    a. Her first birthday _____

    b. Her second birthday _____

    c. Her third birthday _____

    d. Her eighth birthday _____

    e. Her twentieth birthday _____

Uncle Buck$ (cont.)

Directions: Read the text, fill in the chart, and answer the questions below.

2. Hannah's aunt gave Hannah $1 on the day she was born and triples the gift on each birthday until she is 17. Do you think that offer is as generous as Uncle Buck$?

Use your TI-15 and the constant feature (Op1) to find how much she will be given on each birthday. On your TI-15, press [Op1] [×] [3] [Op1]. Press [1] and then [Op1]. Continue to press [Op1] to find the amount she will receive on each birthday. Write the amounts of the gifts on the chart.

Age	$$$	Age	$$$	Age	$$$
0		6		12	
1		7		13	
2		8		14	
3		9		15	
4		10		16	
5		11		17	

Directions: Write the multiplication problem that would give the amount she receives on each of the birthdays below. Give the number of times 3 is used as a factor in each product.

    a. Her first birthday _____

    b. Her second birthday _____

    c. Her third birthday _____

    d. Her eighth birthday _____

    e. Her sixteenth birthday _____

    f. Who was more generous, Hannah's aunt or uncle? _____

Name _____

Dollars and Days

Directions: Follow the steps below.

1. Suppose that you have a dollar. One day later, you have 7 times as much. The second day, you have 7 times the previous day. This continues for 8 days. Use your TI-15 and the ⌃ key to evaluate the amount and complete the chart below.

Day	Factors	Expression	Base	Exponent	Value
1	7	7^1	7	1	7
2	7 x 7				
3		7^3			
4			7		
5				5	
6					117,649
7				7	
8	7 x 7 x 7 x 7 x 7 x 7 x 7 x 7	7^8			

2. Suppose that you have a dollar. One day later, you have 8 times as much. The second day, you have 8 times the previous day. This continues for 8 days. Complete the chart.

Day	Factors	Expression	Base	Exponent	Value
1			8	1	
2					
3					
4					
5				5	
6					
7					
8					

Power Lesson 1

Favorites

Overview

Students will learn to use fractions and decimals to express the same relationships by finding student preferences in several categories and then expressing the results in fractions and decimals.

Mathematics Objective

Understands equivalent forms of basic fractions and decimals and when one form of a number might be more useful than another.

TI-15 Functions

- Fix feature
- Fractions
- Fractions to decimals
- Percent

Materials

- TI-15s
- *Group Response Sheet* activity sheet (page 77; page077.pdf)
- *Tally Sheet* activity sheet (page 78; page078.pdf)
- *Fractions and Decimals* activity sheet (pages 79–80; page079.pdf)
- *The Rest of the Story* activity sheet (page 81; page081.pdf)

Vocabulary

Complete the *Word Wizard* (page 21) vocabulary activity using the words below. Definitions for these words are included on the Teacher Resource CD (glossary.pdf).

- convert
- decimal
- denominator
- fraction
- numerator

Warm-up Activity

1. Have each student choose two whole numbers between 1 and 9 and find their sum.

2. Ask the students to write two fractions, using the numbers they chose for the numerators and their sum as the denominator.

3. Show students how to use the TI-15 to find the sum of the fractions by entering the first numerator, pressing , entering the denominator (sum of the two whole numbers), pressing ⓓ ⊞, entering the second numerator, pressing ⓝ, entering the denominator (sum of the two whole numbers), pressing ⓓ and then Enter.

4. Ask several students to write their sums on the board or overhead.

5. Choose one sum to represent with students. For example, if the sum is $\frac{2}{5} + \frac{3}{5} = 1$, have five students come to the front of the room and place the students into a group of two and a group of three. Ask the class what part of the large group of 5 is represented by each of the small groups.

6. Move the small groups together and ask the students to express the total as a fraction. Remind them that the fraction $\frac{5}{5}$ represents one entire group of 5.

72   #50616—30 Mathematics Lessons Using the TI-15   © Shell Education

Favorites (cont.)

Part One
Explaining the Concept

1. Tell students that they are going to find out the favorites for the class in the following categories: ice cream flavors (I), sports (S), books (B), amusement park rides (R), animals (A), and colors (C). Explain that they will be divided into six groups. Assign each group one category. Each group will survey the class about their favorite items in their category. Each group will need to determine which member will record the results for each rotation.

2. Explain to students that when the groups are paired, each group will survey the other about the favorite item in their category. For example, for the ice cream category, a favorite might be chocolate. Distribute six copies of the *Group Response Sheet* activity sheet (page 77; page077.pdf) to each group. Tell students that they should record all answers for each group and that each response needs to be listed separately on this sheet, even if more than one student in the group has the same response. Tell them that if a student does not have a favorite, they should write down *None*.

3. Allow students time to complete the survey. Tell each group that it will also need to complete a response sheet for its own question.

4. Distribute copies of the *Tally Sheet* activity sheet (page 78; page078.pdf). Instruct students to transfer their data when all groups are finished responding to the questions. Tell them that they will report only the top four responses and the number of votes for the top four responses. If there is a tie among results, have them choose based on alphabetical order. Write the data from the bottom of each group's *Tally Sheet* on the board or overhead.

Applying the Concept

1. Distribute copies of the *Fractions and Decimals* (pages 79–80; page079.pdf) activity sheets. Explain to students that they are going to investigate three ways to represent the class data in each of the categories. Have students write the top 4 results and the numbers on their activity sheets for each category.

2. First students will record the number of votes for the Ice Cream category. Then students need to find the sum of these votes and write it in the *Number* column. Next students will investigate their findings through fractions. Work through the *Fraction* column for the ice cream category with the class. The total in the *Number* column is the numerator of the fractions.

Power Lesson 1

Favorites (cont.)

Part Two
Explaining the Concept

1 Now students will investigate their findings through decimals. Ask students to look at the first fraction in the ice cream category. Tell students to convert a fraction to a decimal, you need to divide the numerator by the denominator. Explain to students how to use long division to convert a fraction to a decimal. Work through the following problem on the board or overhead.

$$\frac{5}{8} = 5 \div 8 = 8\overline{)5.000}$$
$$\phantom{\frac{5}{8} = 5 \div 8 = } \underline{-48}$$
$$\phantom{\frac{5}{8} = 5 \div 8 = 8)}20$$
$$\phantom{\frac{5}{8} = 5 \div 8 = } \underline{-16}$$
$$\phantom{\frac{5}{8} = 5 \div 8 = 8)}40$$
$$\phantom{\frac{5}{8} = 5 \div 8 = } \underline{-40}$$
$$\phantom{\frac{5}{8} = 5 \div 8 = 8)...}0$$

1. Write the decimal and add zero to the dividend.
2. Divide. Multiply the factors. Subtract from the dividend.
3. Add zero to the dividend.
4. Divide. Multiply the factors. Subtract from the dividend.
5. Repeat steps 3–4 until you arrive at 0 or conclude that it is a repeating decimal.

2 Show students two ways the TI-15 can convert fractions to decimals. The first way is to divide the numerator by the denominator using the ÷ key. Have students press [5] [÷] [8] [Enter]. Ask them to round that number to the nearest hundredth.

3 Show them how the TI-15 can round the number for them by using the Fix function. Have students press [Fix] [0.01] [Enter]. The word *Fix* will appear on the screen and the TI-15 will now round all answers to the nearest hundredth. Remind students that they need to hold [On] and [Clear] simultaneously to exit the Fix function.

4 Another way to convert fractions to decimals is to use the [F↔D] key. Have students press [5] [n/d] [8] [Enter]. Then have students press [F↔D]. The TI-15 will convert the fraction to a decimal. If the Fix function is on, the TI-15 will round the decimal to the nearest hundredth.

5 Work through the ice cream category with students. Use long division to convert the first fraction, then use the TI-15 to divide the numerator by the denominator to convert the second fraction, and for the third fraction use the [F↔D] key. Have students vote on which way they would like to convert the fourth fraction in the ice cream category.

Favorites *(cont.)*

Applying the Concept

1 Have students complete the *Fraction* column for each category. Then have students convert the fractions to decimals for all the remaining categories. Have students use long division for the sports category, the ÷ key for the book category, and the F↔D key for the ride category. Tell students they may use whichever method they like for the remaining categories.

2 After students have completed the *Fraction* and *Decimal* columns, have them find fraction and decimal totals for each category. Explain that the total of one makes sense again because you are adding all the parts.

Part Three
Explaining the Concept

1 Distribute copies of *The Rest of the Story* activity sheet (page 81; page081.pdf). Have students use their answers from *Fractions and Decimals* to fill in the total number of votes for the top four in the first column. Ask them to find the fractions for the number of votes for the top four choices out of the number of students in the class.

2 Ask students how to find the fraction for the students whose preferences were not in the top four. On their TI-15s, have them press 1 − the fraction for the top four and then fill in the *Fraction Not in the Top Four* column.

3 Ask students how to find the decimal for the students whose preferences were not in the top four. On their TI-15s, have them subtract 1 − the decimal for the top four and fill in the *Decimal Not in the Top Four* column.

Favorites (cont.)

Applying the Concept

1 After students have finished the chart on *The Rest of the Story* activity sheet (page 81; page081.pdf), have them work in pairs to answer the questions below the chart.

2 When students have finished, come together as a class to discuss the answers. Also investigate how to find the votes not in the top four. Ask students to write a number sentence showing the sum of the top four votes with the votes not in the top four. Now represent these votes as fractions in a number sentence. Ask students, *Why do the fractions total 1?*

Differentiation

- **Below Grade Level—** Allow students to use manipulatives, such as cubes or beans to represent the data collected.

- **Above Grade Level—** Have students write at least two other questions that can be answered looking at the data collected on the *Rest of the Story* activity sheet. Invite some of them to share their questions during the class discussion and answer them as a group.

Extending the Concept

Have students create a graph and chart using all of the data they collected, not just the top four items. Have students then write a paragraph summarizing their conclusions about students' favorites.

Name _____

Group Response Sheet

Directions: In the space below, have a person in your group record the class responses to your question.

       Category: _____

       Recorder: _____

       Group Surveyed: _____

List the favorite of each person in the group being surveyed.

1. _____
2. _____
3. _____
4. _____
5. _____
6. _____

Name _____

Tally Sheet

Directions: Have each student in the group add to this tally sheet for the group he or she surveyed.

Category _____

Add the total number of students from all the sheets. _____

List all the different responses from all the groups. Put a tally mark next to each and then count the total number of tally marks for each response.

List the top four responses and the number of students who gave each of those answers. Give this paper to your teacher.

Response	Number		Response	Number
1.		3.		
2.		4.		

78   #50616—30 Mathematics Lessons Using the TI-15   © Shell Education

Power Lesson 1

Name _____

Fractions and Decimals ¼ 0.25

Directions: Following your teacher's instructions, complete the charts on this page and the next page.

Total Students in Survey _____

	Ice Cream	Number	Fraction	Decimal
1				
2				
3				
4				
Total				

	Sport	Number	Fraction	Decimal
1				
2				
3				
4				
Total				

	Book	Number	Fraction	Decimal
1				
2				
3				
4				
Total				

Fractions and Decimals (cont.)

Directions: Following your teacher's instructions, complete the chart.

	Ride	Number	Fraction	Decimal
1				
2				
3				
4				
Total				

	Animal	Number	Fraction	Decimal
1				
2				
3				
4				
Total				

	Color	Number	Fraction	Decimal
1				
2				
3				
4				
Total				

Name _____

The Rest of the Story

Directions: Complete the chart and answer the questions.

Category	Total Top Four	Fraction Top Four	Decimal Top Four	Fraction Not in the Top Four	Decimal Not in the Top Four
Ice Cream					
Sport					
Book					
Ride					
Animal					
Color					

1. How many students were in the class the day the survey was taken?

   _____

2. How did you find the fraction for the rest of the class?

   _____

3. How did you find the decimal for the rest of the class?

   _____

Lesson 6

How Many Cookies?

Overview

Students will use proportional reasoning to increase or decrease the ingredients in a recipe.

Mathematics Objective

Understands the concepts of ratio and proportion and the relationships between them.

TI-15 Functions

- Constant feature
- Decimals to fractions
- Fractions
- Fractions to decimals
- Operations
- Simplifying fractions

Materials

- TI-15s
- *Sugar Cookies* activity sheet (pages 85–86; page085.pdf)
- *Lots of Cookies* activity sheet (page 87; page087.pdf)

Vocabulary

Complete the *Music Makers* (page 18) vocabulary activity using the words below. Definitions for these words are included on the Teacher Resource CD (glossary.pdf).

- batch
- cup
- dozen
- ratio
- recipe
- tablespoon
- teaspoon

Warm-up Activity

1. Ask each student to write down five numbers between 1 and 10.

2. On their TI-15s, have students press [Opt] [×] [5] and [Opt] again.

3. Have them enter the first number from their lists and press [Opt]. Ask them to write the answer below the first number. They should continue with the other numbers on their lists.

4. Ask students to create fractions from each pair of numbers displayed on the screen after pressing [Opt]. Have them enter the fractions into their TI-15s by putting in the numerator, pressing [n], and then putting in the denominator and [d]. To simplify the fraction, have them press [Simp] and [Enter]. They may need to simplify more than once. Tell them to repeat [Simp] and [Enter] until the number stays the same.

5. After they have entered all the fractions, ask them about their answers. Ask them what it means if you say the ratio of the first number to the second is 1 to 5.

6. Remind students that they need to press and [Clear] at the same time to reset the memory of the TI-15.

#50616—30 Mathematics Lessons Using the TI-15 © Shell Education

Lesson 6

How Many Cookies? (cont.)

Explaining the Concept

1. Ask students: *Have you ever used a recipe? What did you make? What were the ingredients?* Let them share their favorite recipes. Then, ask them what they would do if they needed to provide food for either a large number of people or just a few people.

2. Ask students: *What does a batch of cookies mean?* (*The number of dozens of cookies a recipe makes.*) Ask students: *How many cookies would we need for the whole class if everyone could have four cookies? How many dozens of cookies would we need?*

3. Remind students that they need to divide the number of cookies by 12 to find the number of dozens. Tell them that you only want to consider complete dozens of cookies so for the remaining cookies they will round up to the next dozen.

4. Tell them a recipe makes a batch of 4 dozen cookies. Ask them how many batches they should make for the class. When students divide the number of dozens of cookies by 4, it may come out even. If it does, ask: *"What if you needed 15 dozen cookies?"* Have students divide 15 by 4 on the TI-15. Remind them that they can use the [F↔D] key to change the decimal to a fraction.

5. Show students that pressing [Simp] key and [Enter] will simplify the fraction. Tell them that they should continue pressing the [Simp] key and [Enter] until the answer stays the same. This answer will be in the lowest-terms.

6. Distribute copies of the *Sugar Cookies* activity sheet (pages 85–86; page085.pdf). Explain that the recipe will make a batch of four-dozen cookies.

7. When students finish the activity sheet, ask them if all of the measurements are easy to make or if they need to round some of the measurements. Remind them that cups are measured in thirds, fourths, and halves, and that teaspoons are measured in eighths, fourths, and halves. Remind them that there are three teaspoons in a tablespoon. Ask them how to deal with part of an egg.

© Shell Education   #50616—30 Mathematics Lessons Using the TI-15   **83**

Lesson 6

How Many Cookies? (cont.)

Applying the Concept

1. Distribute copies of the *Lots of Cookies* activity sheet (page 87; page087.pdf) for students to complete. Make sure that they understand how to change each of the measurements into standard kitchen measurements, using the conversions given on the sheet. Show students how to find a fraction of an amount.

2. After reviewing student answers for *Lots of Cookies*, ask how the list of ingredients in question 4 would need to be adjusted if only one dozen cookies were desired. (only $\frac{1}{4}$ of each amount would be needed) Have students use their TI-15s to find the new amount for each ingredient as a simplified fraction or mixed number where appropriate.

3. Have students write a summary of what they have learned. Ask them to consider how ratios can be valuable when cooking.

Differentiation

- **Below Grade Level**—Make paper models of each of the measures in the recipe for *Lots of Cookies*. Have students group the number in the recipe and the number required.

- **Above Grade Level**—Have the students investigate how to change the recipe in *Lots of Cookies* to a normal recipe all in one calculation. Ask them to write a general description that could be changed for any number of dozens of cookies.

Extending the Concept

- Discuss the concept of changing units of measurement. Ask them how to change from a smaller unit of measure to a larger, e.g., inches to yards.

- Ask them if they thought the units of measure used in the kitchen were confusing. Look at a recipe expressed in the metric system and the conversions between metric measurements.

Name _____

Lesson 6

Sugar Cookies

Directions: Use the recipe in the box and your TI-15 to answer the questions.

> This recipe will make a batch of four dozen cookies.
>
> $2\frac{1}{2}$ cups all-purpose flour   1 cup butter, softened
>
> 1 teaspoon baking soda $1\frac{1}{2}$ cups white sugar
>
> $\frac{1}{2}$ teaspoon baking powder   1 egg
>
> 1 teaspoon vanilla extract

1. Sofia needs to bake cookies for 48 people. She wants to give each person 2 cookies.

   a. How many cookies does she need? _____

   b. How many dozen cookies does she need? _____

   c. The number of cookies a recipe makes is called a batch of cookies. How many batches does Sofia need?

   _____

How much of each ingredient will she need to make two batches of cookies? Use your TI-15 and follow the steps below to double the recipe.

   Step 1   Press [Opl] [×] [2] [Opl].
   Step 2   To find out how much flour she needs, press [2] [Unit] [1] [n] [2] [d].
   Step 3   Press [Opl].

   d. How much flour does she need? _____

   e. How much baking soda does she need? _____

   f. How much baking powder does she need? _____

   g. How much butter does she need? _____

   h. How much sugar does she need? _____

   i. How much vanilla does she need? _____

   j. How many eggs does she need? _____

Sugar Cookies (cont.)

Directions: Use the recipe in the box to answer the questions.

> This recipe will make a batch of four dozen sugar cookies.
>
> $2\frac{1}{2}$ cups all-purpose flour   1 cup butter, softened
>
> 1 teaspoon baking soda $1\frac{1}{2}$ cups white sugar
>
> $\frac{1}{2}$ teaspoon baking powder   1 egg
>
> 1 teaspoon vanilla extract

2. Sofia needs to bake cookies for 108 people. She wants to give each person 2 cookies.

   a. How many cookies will she need? _____

   b. How many dozens is this? _____

   c. How many batches does she need? _____

Store this number in Op2 by pressing [Op2] [×] and then enter the number of batches. Press [Op2] again. Calculate the amount of ingredients needed to make the batches of cookies. Enter the measurement listed in the recipe and then press [Op2]. Remember, to enter mixed fractions you need to use the [Unit] [n] and [d] keys.

   d. How much flour does she need? _____

   e. How much baking soda does she need? _____

   f. How much baking powder does she need? _____

   g. How much butter does she need? _____

   h. How much sugar does she need? _____

   i. How much vanilla does she need? _____

   j. How many eggs does she need? What would you do about the eggs? _____

Name _____

Lesson 6

Lots of Cookies

Directions: Use the recipe in the box to answer the questions.

This recipe will make 384 sugar cookies.		
10 lb. flour	$\frac{1}{2}$ cup vanilla extract	4 lb. sugar
1 cup milk	8 lb. butter	

Nadia wants to use the school cafeteria's sugar-cookie recipe. When she asked the cafeteria manager for the recipe, she was given the recipe above. Nadia knows that she does not want to make that many cookies. She also knows that her kitchen equipment measures in cups, tablespoons, and teaspoons. Nadia did some research and found these facts.

- One pound of flour contains about 4 cups of flour.
- One pound of sugar contains about 2 cups of sugar.
- One pound of butter contains about 2 cups of butter.
- One cup of liquid contains 16 tablespoons of liquid.

1. Change the flour, butter, and sugar to cups and the vanilla extract and milk to tablespoons. Write your answers on the lines below.

    - flour: _____
    - sugar: _____
    - butter: _____
    - vanilla extract: _____
    - milk: _____

2. How many dozen cookies would this recipe make? _____

3. Nadia wants to make four-dozen cookies. What fraction of this recipe should she make?

    _____

4. Use this number as a fraction in [Opl] and find the amount of each ingredient Nadia needs. Write Nadia's new recipe below. Use the converted amounts listed in problem 1.

    - _____ cups of flour
    - _____ cups of sugar
    - _____ cups of butter
    - _____ tablespoons of vanilla extract
    - _____ tablespoons of milk

© Shell Education
#50616—30 Mathematics Lessons Using the TI-15

Lesson 7

Fill in the Blank

Overview

Students will use their TI-15s to solve for variables.

Mathematics Objective

Knows that a variable is a letter or symbol that stands for one or more numbers.

TI-15 Functions

- Operations
- Problem solving

Materials

- TI-15s
- *Missing Numbers* activity sheet (pages 91–92; page091.pdf)
- *Write Your Own* activity sheet (page 93; page093.pdf)

Vocabulary

Complete the *Which Statement Is Accurate?* (page 17) vocabulary activity using the words below. Definitions for these words are included on the Teacher Resource CD (glossary.pdf).

- addend
- addition
- difference
- inequality
- minuend
- solution
- subtraction
- subtrahend
- sum
- variable

Warm-up Activity

1. Write an addition equation on the board or overhead. Depending on your students, there can be multiple addends or the addends can be four or more digits in length. Have students identify the addends.

2. Instruct students to use their TI-15s to find the sum. Repeat a few times with other addition equations.

3. Write a subtraction equation on the board. Depending on your students, there can be multiple subtrahends, or the numbers can be four or more digits in length. Have students identify the minuend and the subtrahend.

4. Instruct students to use their TI-15s to find the difference. Repeat a few times with other subtraction equations.

Fill in the Blank (cont.)

Explaining the Concept

1 Write an addition equation on the board or overhead, covering the sum before showing the equation to the students. Ask students to guess the hidden number before showing them the answer. Then write another addition equation but this time hide the second addend. Ask students to guess the second addend before showing it to them.

2 Tell students they are going to learn how to use the Problem Solving feature on the TI-15. Have them hold (On) and (Clear) at the same time for a few seconds to get the TI-15 ready to accept a new function. To put the TI-15 into the Problem Solving mode, tell the students to press ◈ (Mode). The display will flash YES and NO momentarily and then automatically change to AUTO and MAN. The line under AUTO means that that choice is active. This means that the TI-15 will automatically create and display problems.

3 Have students press ◈ to select the level of difficulty. The 1 means that the lowest level of difficulty has been chosen. If you want to change the level, use the ◈ or ◈ to move among the choices.

4 Ask students to press ◈ again to select the operation. The + means that addition has been selected. If you want to change the operation, use the ◈ or ◈ to move among the choices. Press (Clear).

5 The TI-15 will display an addition equation with a ? in either the position of the second addend or the sum. This is known as the variable. Ask student volunteers to write their equations on the board or overhead. Have the class give the value of the variable for each. Tell them that the number is a *solution* of the equation. Have all students enter their solutions and press (Enter).

6 Tell students if the display says YES, they entered the correct addend and that the TI-15 will give them new equations. Explain that NO means that their solutions are not correct and the TI-15 will display inequality signs to show them if their answers are too large or too small before it displays the problems again. The correct answer is provided after three incorrect attempts to find a solution. Have students experiment with wrong answers to see how the TI-15 responds.

7 Have students complete a total of five equations and observe that their TI-15s display scores for the five items. Tell the students to press ◈ to exit the Problem Solving mode.

8 Distribute copies of *Missing Numbers* (pages 91–92; page091.pdf). Discuss questions 2 and 3 after students have completed the first two columns of the first chart. Then have students complete the rest of the activity sheet.

Lesson 7

Fill in the Blank (cont.)

Differentiation

- **Below Grade Level—** Have students work in pairs and use chips or blocks to illustrate the parts of the equation. Show them that for an addition equation, they put all the chips together to form a sum and separate them to find an addend. For a subtraction problem, they separate the chips to find a difference and put them together to find a minuend.

- **Above Grade Level—** With the activity sheet, *Write Your Own*, have students explore the manual problem solving function using more than one question mark in their equations.

Applying the Concept

1. Distribute copies of the *Write Your Own* activity sheet (page 93; page093.pdf). Help students follow the steps to put their TI-15s into Manual Problem-Solving mode. Use ? + ? = 3 to show them that when multiple question marks are used, multiple solutions are possible.

2. After students have completed the activity sheet, discuss questions 1–3. Help them see that they should subtract when the question mark is after the equal sign; they should subtract the difference from the minuend when the question mark is in the location of the subtrahend; and they should add the subtrahend and the difference when the missing number is the minuend.

3. Ask students to share their word problems. Ask them to explain what their solutions mean, for instance, *Pedro had 11 cookies when he started*.

Extending the Concept

Have students try problem solving with multiplication and division, looking at factors, products, divisors, dividends, and quotients. Ask them which operation behaves in a manner similar to addition and which operation is similar to subtraction. (*Addition is similar to multiplication and subtraction is similar to division.*)

Name _____

Missing Numbers

Directions: Follow the steps below on your TI-15.

Step 1 Press (On) and (Clear) at the same time for a few seconds. This will clear the TI-15.

Step 2 Press (Clear).

Step 3 Press ◈ and then (Mode). Wait until you see AUTO and MAN.

Step 4 Press ◈ to select the level of difficulty.

Step 5 Press ◈ to place the line under 2 and press [Enter].

Step 6 Press ◈ again to select the operation.

Step 7 Place the line under the addition sign (+) and press [Enter].

Step 8 Press (Clear). The display should show an equation with a question mark. Write the equation in the chart below. Figure out the number that belongs in place of the question mark. Type in that number and press [Enter].

Step 9 If your answer is correct, you will see YES and the TI-15 will display another problem. If your answer is incorrect, the display will show NO and show the problem again. It will allow you to try three times, and then show the correct answer. After five problems, the TI-15 will show your score with the number for YES and NO.

Directions: Complete the chart below for five completed problems. Leave the *How* column blank for now.

Equation	? (Solution)	How

© Shell Education

Lesson 7

Missing Numbers (cont.)

Directions: Answer the following questions.

1. How many equations did you get right? _____

2. Some of the questions had the question mark *after* the equal sign, such as 50 + 60 = **?**. How did you find the answer to these items? Write this answer in the *How* column on the first page.

3. Some of the questions had the question mark *in front* of the equal sign, such as 9 + **?** = 809. How did you find the answer to these items? Write this answer in the *How* column on the first page.

4. Complete five more equations. On the chart below, fill out the *How* column first and put your answers in the *?(Solution)* column before entering your values. If the answer is not correct, fix your answers in the columns and then try again.

Equation	? (Solution)	How

Name _____

Write Your Own

Directions: Follow the steps below.

You can write your own equations to solve on your TI-15.

Step 1 Press ◈ and then (Mode).

Step 2 Press ◁▷ to move the cursor so that MAN is underlined.

Step 3 Press [Enter] and then (Mode).

Step 4 Press [2] [3] [−] [2] [Enter] [?] [Enter]. The TI-15 displays 1 SOL. This means that the problem has only one solution.

Step 5 Press [2] [1] [Enter]. The display shows YES.

Use the TI-15 to find the value for the question marks for the equations below. In the *How* column, tell which numbers you would add or subtract.

Equation	? (Solution)	How
305 − 5 = ?		
180 − ? = 80		
? − 120 = 10		
161 − 11 = ?		
190 − 17 = ?		
? − 88 = 120		

Directions: Make up subtraction equations with the question mark in the given place. Tell whether you will add or subtract to find the answer. Check your answer on your TI-15.

1. The question mark is after the equal sign _____

2. The question mark is after the minus sign _____

3. The question mark is at the beginning of the equation _____

Directions: On a separate sheet of paper, write a word problem for each equation you wrote in numbers 1–3. For example, for the equation ? − 15 = 25, you might write: *Pedro had some cookies and gave 15 of them away. He had 25 left. How many cookies did he start with?*

Lesson 8

Moving Along

Overview

Students will learn to create and interpret arithmetic sequences.

Mathematics Objective

Recognizes a wide variety of patterns and the rules that explain them.

TI-15 Functions

- Constant feature
- Operations

Materials

- TI-15s
- *Snail Mail* activity sheet (pages 97–98; page097.pdf)
- *Hikers* activity sheet (page 99; page099.pdf)

Vocabulary

Complete the *Sharing Mathematics* (page 16) vocabulary activity using the words below. Definitions for these words are included on the Teacher Resource CD (glossary.pdf).

- common difference
- difference
- finite
- infinite
- sequence
- term

Warm-up Activity

1. Tell students to press [Op1] [+] [3] [Op1] on their TI-15s.

2. Have them press [4] [3] [Op1]. Ask students what happened. (*3 was added to 43 to equal 46.*)

3. Point out to students the "counter" that appears at the lower left of the display. Ask students what number they think the answer will be when the counter reaches 10. Have them write down their guesses and then press [Op1] nine more times.

4. Ask students if it they can predict what the answer will be after they press [Op1] 30 times. Have them write down their guesses before they press [Op1] an additional 20 times to bring the counter to 30.

5. Ask them if it they can predict what the answer will be after they press [Op1] 50 times. Have them write down their guesses before they press [Op1] an additional 30 times to bring the counter to 50.

6. Remind students how to clear what is stored in the Constant feature (Op1) by pressing and (Clear) at the same time.

Lesson 8

Moving Along (cont.)

Explaining the Concept

1. Ask the class to discuss situations where the same number would be added over and over. They might mention a car on cruise control where the number of miles traveled each hour would be the same, or saving the same amount from your allowance each week.

2. Tell students that they are going to investigate what happens when the same amount is added repeatedly. Distribute copies of the *Snail Mail* activity sheet (pages 97–98; page097.pdf) to students, and work through the first problem with students.

3. Explain that they are setting up a pattern that is a sequence. This means that it is possible to describe in words how the next number will be determined, if you know the beginning value. Ask them to describe how they would find the eleventh number in this sequence. (*Adding fifteen to the previous number.*) Tell students that each number in the sequence is called a *term*.

4. Tell students that the amount added to each term is called the *common difference*. Explain that this kind of sequence is called an *arithmetic sequence*. Tell them that in an arithmetic sequence, the same number is added or subtracted to create the next term.

5. Explain that it is possible to use [Opl] to repeat the addition. Remind students that the counter shows how many times [Opl] has been pressed.

6. Have students complete problems 2 and 3 on the activity sheet. Have them describe the beginning value and the common difference of the sequences. Ask them to describe the sequences for the brown and gray snails. Ask them which snail is moving faster and how they can tell. (*The gray snail is moving faster because it moves farther each day.*)

7. Ask students how far the sequences could be continued. They should realize that the sequences continue forever (infinite), but the portions describing the snail's movements would just represent a part of it (finite).

Lesson 8

Moving Along (cont.)

Differentiation

- **Below Grade Level**—For the activity sheet *Hikers*, have students follow the hikers' progress on tape measures. They should mark the distance traveled each day using sticky notes. Have them use the sticky notes to find each day's new position and to illustrate the amount that is added each day.

- **Above Grade Level**—Give students the opportunity to write a sequence on one side of a note card and then identify only two terms on the other side of that note card. Have them trade with a partner. Have each student use the side with only two terms to determine the common difference and the first five terms of the sequence. Then have them write the expression that corresponds with their given sequence.

Applying the Concept

1 Show students the chart below. Tell them that the bottom line represents part of an arithmetic sequence. Ask them to identify the common difference. (5) Ask them how to move forward and backward to complete the chart.

1	2	3	4	5	6	7	8
			18	23			

2 Distribute copies of the *Hikers* activity sheet (page 99; page099.pdf) to students and have them complete their sheets.

Extending the Concept

- Help students write expressions to represent patterns in the tables for both *Snail Mail* and *Hikers*.

- Show students that a sequence is often written, for example, 3, 7, 11, 15, ..., and the term numbers are understood. Tell them that the three dots at the end are called an *ellipsis* and indicate that the pattern continues forever. Have them write the sequences from the two activity sheets, using this notation.

- Ask students to identify the common difference and to find the remaining positive numbers in the arithmetic sequence 17, 14, 11, ... Discuss that the common difference is subtracted and ask students to speculate on the continuation of this sequence.

Name _____

Snail Mail

Directions: Using the information given, answer the questions below.

1. Tamara went to her mailbox and saw a snail crawling up the post of the mailbox. The snail was 42 mm from the bottom of the post. When she went to get the mail the next day, the snail was 57 mm from the bottom of the post. On the second day, the snail was 72 mm from the bottom of the post.

   a. What is the total distance the snail had moved in the two days? _____

   b. Suppose the snail continues to move up the post at the same rate. How far would it move each day? Fill in the information for Days 3 through 10 on the chart below.

Day	0	1	2	3	4	5	6	7	8	9	10
Position (mm)	42	57	72								

The difference between each day's position and the next day's position is called the *common difference*. Use your TI-15 and follow the steps below to complete the chart through Day 10.

   Step 1   Press [On] and [Clear] at the same time to reset the TI-15. Press [Clear].

   Step 2   Press [Opl] [+] [1] [5].

   Step 3   Press [Opl].

   Step 4   Enter the snail's position on day zero.

   Step 5   Press [Opl].

   c. What does the TI-15 show? _____

   Step 6   Press [Opl] again.

   d. What does the TI-15 show? _____

   Step 7   Press [Opl] again.

   e. What does the TI-15 show? _____

   f. Notice that the TI-15 keeps count of how many times you have pressed [Opl]. Continue to press [Opl] until you reach Day 10. Fill in the chart above as you go.

Lesson 8

Snail Mail (cont.)

Directions: Using the information given, answer the questions below.

2. The snail disappeared after Day 10, but a few weeks later, Tamara saw another snail. She recorded the snail's progress over three days. It is in the chart below.

Day	0	1	2	3	4	5	6	7	8	9	10
Position (mm)	37	60	83	106							

   a. How much did the distance increase each day? _____

   b. Which snail moved faster, the first one or the second one? _____

   c. How would you use the [Op1] key to set up the addition for the snail's movement?
   _____

   d. Fill out the chart above for the rest of the days.

3. A few weeks later there were two snails on the post. The gray snail was 53 mm from the bottom of the post. The brown snail was 75 mm from the bottom of the post. The next day, the gray snail was 80 mm from the bottom of the post, and the brown snail was 98 mm from the bottom of the post. They continued to move at the same rate.

   a. How far is the gray snail moving each day? _____

   b. How will you find its position from one day to the next? _____

   c. How far is the brown snail moving each day? _____

   d. How will you find its position from one day to the next? _____

   e. Fill out the chart below for the snails.

Day	0	1	2	3	4	5	6	7	8	9	10
Gray Position (mm)											
Brown Position (mm)											

   f. Which day(s) will the gray snail be closest to the brown snail? _____

Name _____

Hikers

Directions: Complete the problems below.

1. Jermaine's Boy Scout troop went on a hike. The boys' parents drove them to the starting point and picked them up at the end. They hiked for 8 days and traveled the same number of miles every day. On Day 5, they were 209 miles from home. On Day 6, they were 243 miles from home.

   a. How far did they hike each day? _____

   b. Complete the chart below.

Day	0	1	2	3	4	5	6	7	8
Position (miles)									

   c. How far from home did they start hiking? _____

   d. What are the keystrokes for using your TI-15 to create the table?
   _____

   e. If you were given only Day 0 and Day 8 information, how could you have determined how far they hiked each day?
   _____

2. The troop is planning another hike. The parents will drop them off 50 miles from home and pick them up 210 miles from home. The trip will take five days.

   a. Fill in the beginning and end of the trip on the chart below.

   b. How many miles should they hike each day? _____

   c. Fill in the rest of the days on the chart below.

   d. Use words to tell someone else how to find where the troop is each day.
   _____
   _____

Day	0	1	2	3	4	5
Position (miles)						

Lesson 9

Order Matters

Overview

Students will learn to apply the algebraic order of operations.

Mathematics Objective

Uses basic and advanced procedures while performing the processes of computation.

TI-15 Function

- Operations

Materials

- TI-15s
- *Which Way Did They Go?* activity sheet (pages 103–104; page103.pdf)
- *All Mixed Up* activity sheet (page 105; page105.pdf)

Vocabulary

Complete the *Chart and Match* (page 19) vocabulary activity using the words below. Definitions for these words are included on the Teacher Resource CD (glossary.pdf).

- grouping
- order of operations
- unique

Warm-up Activity

1. Ask students to find the answers to the following problems without using their TI-15s. Do not give them any hints on the correct order of operations that should be used to complete the computations.

   a. 5 + 6 x 2 *(17)*　　**b.** 6 x 2 + 5 *(17)*　　**c.** 15 – 3 x 4 *(3)*

   d. 7 + 2 x 3 – 1 *(12)*　　**e.** 5 x 3 – 2 x 3 *(9)*

2. List all of the answers that students got for each problem. There are likely to be different answers depending on the order of operations the students used. Have volunteers explain how they got their answers.

3. Demonstrate how to determine the correct value for each expression. Then have students test if the TI-15 follows the order of operations correctly. Tell them to enter each problem in one string and press [Enter] only once at the end.

4. Have students compare the answers on the TI-15 to the correct answers. Have students draw a conclusion as to whether the TI-15 follows the order of operations.

Order Matters (cont.)

Explaining the Concept

1 Ask the class to define the word *unique*. They should know that something that is unique is one of a kind. Ask them for some events or objects that would be considered unique. Explain that in mathematics the word *unique* means "just one." As an example, tell them that the sum of 7 + 5 has a unique answer of 12. Ask students to discuss why it is important that the problems in the warm-up activity should have unique answers.

2 Explain that there is an order that must be followed when simplifying expressions. Any multiplication and division must be done from left to right before any addition or subtraction is done from left to right. Write the example, 8 + 12 ÷ 4 x 5. Explain each step as you underline 12 ÷ 4, and underneath write the next step of 8 + 3 x 5. Next, underline 3 x 5 and underneath write 8 + 15. Finally, underline 8 + 15 and underneath write 23. Have students enter the original expression to find the simplified value of 23. Tell them that the TI-15 and most scientific calculators have the correct order of operations built into them, but that most 4-function calculators (basic addition, subtraction, multiplication, and division only) do not. If you have 4-function calculators available, have students repeat the warm-up problems on them. Show that if such a calculator were used, the problem would be simplified only from left to right resulting in an incorrect value of 25.

3 Tell students that they are going to use the TI-15s to help them learn the correct order of operations. Distribute copies of the *Which Way Did They Go?* activity sheet (pages 103–104; page103.pdf) to students.

4 Discuss that the order of operations is often remembered by using the acronym MDAS, standing for Multiplication, Division, Addition, and Subtraction. Tell students that one way to remember the acronym is to use "My Dear Aunt Sally." But remind students that multiplication and division are considered on the same level and must be calculated from left to right. Also, addition and subtraction are considered on the same level and must be calculated from left to right.

5 Work as a class to answer questions 1–5 on the first page of the *Which Way Did They Go?* activity sheet. Then have students practice using the order of operations by solving problems 6–11. Go over the answers as a class.

6 Tell students that they will now use the order of operations to find the missing number. Have students complete *Which Way Did They Go?* using MDAS. Help them get started by working questions 12 and 20 with the class. Have them show their work such as ? + 14 = 19 as the middle step of problem 12.

Lesson 9

Order Matters (cont.)

Applying the Concept

1 Distribute copies of the *All Mixed Up* activity sheet (page 105; page105.pdf). Ask students to propose possible methods to find the answer to the first problem using their TI-15s. Then, have them use their TI-15s to find the answer.

2 Have students complete problems 1–9. Make sure they realize that addition and subtraction are performed from left to right and that addition does not necessarily precede subtraction.

3 Have students complete problems 10–12. Make sure they see the left to right order in these problems, and tell them that multiplication does not necessarily precede division.

4 Have students complete problems 11–15. Then have them go back and identify the MDAS order in problems 10–12.

Differentiation

- **Below Grade Level—** Have students use markers to underline the order of operations. Use red for the first operation, blue for the second, green for the third, and orange for the fourth. Then, have them write the problems they will do in the same colors.

- **Above Grade Level—** Ask students who are familiar with exponents to use their TI-15s to evaluate expressions like: $75 - 3^2 \times 2$. Tell students that *E* stands for exponents. Ask them where they should put the E in the MDAS acronym. *(at the beginning)*

Extending the Concept

Explain to students that it is sometimes necessary to compute in a way that requires another level to the order of operations. For example, suppose that on Monday, Julio ran 3 miles in the morning and 4 miles in the afternoon. On Tuesday, he ran twice as far as he did on Monday. To show that 3 and 4 should be added first, they should use parentheses and write 2 (3 + 4). Have students write and evaluate expressions with parentheses using their TI-15s. Ask them how they could have evaluated the expression without the use of parentheses. (2 x 3 + 2 x 4) Show them that this is an example of the distributive property.

Name _____

Which Way Did They Go?

Directions: Answer the following questions.

1. In a problem that has multiplication and addition, which operation will come first?

   _____

2. In a problem that has multiplication and subtraction, which operation will come first?

   _____

3. In a problem that has division and addition, which operation will come first?

   _____

4. In a problem that has multiplication and division, which operation will come first?

   _____

5. What does MDAS stand for?

   _____

Directions: Find the answers to the problems using MDAS. Do not use your TI-15. Show your work.

6. $20 - 5 \times 2 =$ _____

7. $30 \div 3 + 2 =$ _____

8. $16 + 8 \times 3 =$ _____

9. $40 - 10 \div 5 =$ _____

10. $15 \times 3 - 2 =$ _____

11. $6 \times 7 + 3 =$ _____

Which Way Did They Go? (cont.)

Directions: Complete each problem. Show your work. Then use your TI-15 to check your answers.

12. ? + 7 x 2 = 19 _____

13. ? + 9 x 3 = 35 _____

14. 8 x 2 + ? = 20 _____

15. ? − 2 x 3 = 5 _____

16. ? − 4 x 2 = 7 _____

17. 3 x 5 − ? = 11 _____

18. ? + 12 ÷ 4 = 11 _____

19. ? + 15 ÷ 5 = 23 _____

20. 28 ÷ 2 + ? = 19 _____

21. ? − 10 ÷ 5 = 38 _____

22. ? − 16 ÷ 4 = 28 _____

23. 42 ÷ 6 + ? = 15 _____

24. 100 ÷ 10 + ? = 15 _____

25. ? − 100 ÷ 4 = 175 _____

Name _____

All Mixed Up

Directions: Use your TI-15 to find the answer to each problem. Show the work that would produce that answer. Remember MDAS.

1. 15 – 2 + 8 = _____

2. 30 ÷ 3 x 5 = _____

3. 12 – 4 x 1 + 6 = _____

4. 20 ÷ 2 – 3 x 2 = _____

5. 5 x 10 ÷ 5 – 6 + 1 = _____

Directions: Complete each problem. Show your work. Then use your TI-15 to check your answers.

6. 15 – 8 – ? = 5 _____

7. ? + 3 x 5 = 35 _____

8. 12 – 4 x 2 + ? = 10 _____

9. ? – 28 ÷ 4 = 4 _____

10. 30 ÷ 3 + ? = 18 _____

11. 6 ÷ 2 + ? = 46 _____

12. 40 + 8 ÷ 2 – ? = 40 _____

13. 36 + ? – 3 x 7 = 19 _____

14. ? + 12 ÷ 3 – 1 = 17 _____

15. 50 x 2 + ? – 5 = 110 _____

Lesson 10

The Shapes of Numbers

Overview

Students will use their TI-15s to investigate patterns that produce triangular numbers, square numbers, and Fibonacci numbers.

Mathematics Objective

Recognizes a wide variety of patterns and the rules that explain them.

TI-15 Functions

- Operations
- Repeated addition

Materials

- TI-15s
- *Dots* activity sheet (pages 109–110; page109.pdf)
- *The Italian Connection* activity sheet (page 111; page111.pdf)

 Vocabulary

Complete the *Frayer Model* (page 20) vocabulary activity using the words below. Definitions for these words are included on the Teacher Resource CD (glossary.pdf).

- ellipsis
- Fibonacci sequence
- infinite
- pattern
- sequence
- square number
- term
- triangular number

Warm-up Activity

1. Write the chart below on the board or overhead.

Term	1	2	3	4	5	6	7	8	9
Value									

2. Ask a student to pick a number between 1 and 10.

3. Tell students to enter the number in their TI-15s and press [Enter]. Write the number beneath the 1 on the chart.

4. Ask another student to pick a different number between 1 and 10.

5. Have students press [+], then have them enter the number the second student chose, and [Enter]. Write the number beneath the 2 on the chart.

6. Have students continue to push [+], the number the second student chose, and [Enter], filling in each result on the chart.

The Shapes of Numbers (cont.)

Explaining the Concept

1 Ask students to describe the pattern they created in the warm-up activity. They should describe it as starting with a number and adding the same number repeatedly. Remind students that the pattern is called a *sequence* and the values in it are called *terms*. We keep track of the terms by numbering them. Ask students to identify the value of the eighth term. Ask students how many terms the sequence could have. They should realize that the pattern goes on forever. Remind them that this is an *infinite* sequence because it can continue forever.

2 Ask students if they could create a sequence by multiplying by the same number each time. Try it by having the students use the same beginning number they used in the warm-up activity. They should press ⊠ and the second number repeatedly. Write the new sequence on the board or overhead. Ask the students if this is also an infinite sequence.

3 Tell students that they are going to look at sequences that are created in other ways, with patterns that can be described in different ways. Distribute copies of the *Dots* activity sheet (pages 109–110; page109.pdf). Tell students that they are going to look for numeric patterns from geometric patterns of dots.

4 In problem 1, have students identify the term number as the number of dots in the bottom row and the term value as the total number of dots. Point out that the number of dots in the bottom row equals the number of rows.

5 To complete problem 2, help them identify the number to be added on each step. This will be the same as the term number—the number of rows.

6 After students have completed the chart, ask them: *Was the same number added to each term?* (It was not.) Ask: *Is it always possible to predict the next number to be added?* (It is always the term number or each time you add one more than you did the last time.)

7 Ask students why these are called *triangular numbers*. Have them draw the triangles on the sets of dots. The rest of the *Dots* activity sheet creates a pattern of square numbers. After students have completed problem 3, ask them to identify the term number from the dot pattern (the number of dots in the bottom row or the number of rows) and the value of the terms (the total number of dots in the figure).

8 Make sure that students recognize that they are adding successive odd numbers to the list to create the next number. Then, have students draw the squares on the dot patterns to emphasize the concept of *square* numbers.

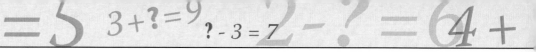

Lesson 10

The Shapes of Numbers (cont.)

Differentiation

- **Below Grade Level**—For the *Dots* activity sheet, have students work in pairs and use chips to create the patterns of dots, counting the number of rows and chips as they arrange them.

- **Above Grade Level**— Ask students to determine the 10th, 20th, and 30th term if the chart on page 110 were to continue. Have students investigate the use of a word square to indicate a number raised to the second power. Have them write what would belong in the "value" box of page 110 if the "term" box showed n. (n^2)

Applying the Concept

1 Distribute copies of *The Italian Connection* activity sheet (page 111; page111.pdf). Each new term of the sequence is generated by adding the previous two terms together: 1, 1, 2, 3, 5, 8, 13, 21, 34, …

2 Show students how to use an *ellipsis* (the three dots) to indicate that the sequence continues forever. Have them complete their activity sheets.

3 After students have completed their activity sheets, briefly discuss Fibonacci and his famous sequence as a class.

Extending the Concept

- Have students write the sequences on their *Dots* activity sheets using an ellipsis.

- Have them investigate the occurrence of Fibonacci numbers in nature.

Name _____

Dots

Lesson 10

Directions: Follow the steps below.

1. Look at the pattern of dots. Under each set on the top line, write the number of dots on the bottom row. On the lower line, write the total number of dots in each set.

   Set 1   Set 2   Set 3   Set 4

   Term   _____   _____   _____   _____

   Value   _____   _____   _____   _____

   a. How many rows are in the first set of dots in the pattern? _____

   b. How many rows are in the second set of dots in the pattern? _____

   c. How many rows are in the third set of dots in the pattern? _____

   d. How many rows are in the fourth set of dots in the pattern? _____

   e. How would you draw the next set of dots in the pattern? Draw it in the space below.

2. Use your TI-15 and follow the steps below to find the total number of dots in each set of dots in the pattern.

   Step 1   Press [1] [Enter]. Write 1 on the chart under the 1.
   Step 2   Press [+] [2] [Enter]. Write the new answer on the chart under the 2.
   Step 3   Press [+] [3] [Enter]. Write the new answer on the chart under the 3.

   Continue until the chart is complete.

Rows	1	2	3	4	5	6	7	8	9
Dots									

© Shell Education   #50616—30 Mathematics Lessons Using the TI-15   **109**

Lesson 10

Dots (cont.)

Directions: Follow the steps below.

3. Look at the pattern of dots. Under each set of dots, write the number of dots in the bottom row. On the lower line, write the total number of dots in each set of dots.

   Set 1   Set 2   Set 3   Set 4   Set 5

Term _____ _____ _____ _____ _____

Value _____ _____ _____ _____ _____

   a. How many more dots does the second set have than the first set? _____

   b. How many more dots does the third set have than the second set? _____

   c. How many more dots does the fourth set have than the third set? _____

   d. How many more dots does the fifth set have than the fourth set? _____

4. Use your TI-15 and follow the steps below to figure out the total number of dots in each set.

   Step 1 Press [1] [Enter]. Write 1 on the chart under the 1.
   Step 2 Press [+] [3] [Enter]. Write the new answer on the chart under the 2.
   Step 3 Press [+] [5] [Enter]. Write the new answer on the chart under the 3.

Continue until the chart is completed.

Term	1	2	3	4	5	6	7	8	9
Value									

#50616—30 Mathematics Lessons Using the TI-15 © Shell Education

Name _____

The Italian Connection

Directions: Use your TI-15 and follow the steps below. Keep track of your answers in the chart below.

Step 1 Press [1] [Enter]. Fill in 1 as the value of the first term.

Step 2 Press [+] [0] [Enter]. Fill in the answer as the value of the second term.

Step 3 Press [+] [1] [Enter]. You are using 1 because it is the answer to the last sum.

At this point, the display should show 1 + 1 = 2.

Step 4 Continue to add numbers to each sum. The number added will always be the previous term. This will always be the first number in the equation on the display. In this case, it is 1. So, press [+] [1] [Enter].

Step 5 The display shows 2 + 1 = 3, so press [+] [2] [Enter], because 2 was the previous sum and is the first number on the display.

Step 6 The next addend should be 3 because 3 was the previous term. It is also the first number shown on the display in 3 + 2 = 5. Continue to compute the sums until the chart is finished.

Term	1	2	3	4	5	6	7	8	9
Value									

This sequence of patterns is called the *Fibonacci Sequence*. It was named after the man who discovered it. Fibonacci was a mathematician who lived during the Middle Ages in Pisa, Italy. He used this number sequence to model the growth of rabbit populations.

Now write a rule in your own words to describe the Fibonacci Sequence.

_____
_____
_____
_____

© Shell Education

Power Lesson 2

A New Look

Overview

Students will learn to use patterns to create tables and graph them. They will learn to interpret graphs and use them to predict values.

Mathematics Objective

Knows basic characteristics and features of the rectangular coordinate system.

Understands various representations of patterns and functions and the relationships among them.

TI-15 Functions

- Constant feature
- Operations

Materials

- TI-15s
- graph paper
- transparency of graph paper
- *Dozens and Dozens* activity sheet (pages 117–118; page117.pdf)
- *The Cookie Business* activity sheet (pages 119–121; page119.pdf)

Vocabulary

Complete the *Word Wizard* (page 21) vocabulary activity using the words below. Definitions for these words are included on the Teacher Resource CD (glossary.pdf).

- break even
- expenses
- horizontal axis (*x*-axis)
- loss
- profit
- scale
- vertical axis (*y*-axis)

Warm-up Activity

1. To clear any number stored in the Constant feature (Op1), have students press ⓞ and ⓒ keys at the same time.

2. Have each student pick a whole number between 2 and 5.

3. Instruct the students to press [Op1], the multiplication sign, the number chosen, and [Op1] again.

4. Ask them to press each of the whole numbers between 1 and 10 followed by [Op1]. Have them make a list of the answers from each operation.

5. Record students answers in a table or T-chart.

6. Ask students how each answer compares to the previous number.

A New Look (cont.)

Part One
Explaining the Concept

1 Draw the table below on the board or overhead. Explain to students that you are planning on selling some old CDs. The table represents the amount of money you will make based on the number of CDs you sell. Ask students what the cost of each CD is. ($2.00)

Number of CDs	1	2	3	4	5	6	7
Sales ($)	2	4	6	8	10	12	14

2 Tell students that when a table is graphed, in most cases, the top row (or left column) represents the independent values (what you choose) and is used for the horizontal axis (*x*-axis) and the lower row (or right column) represents the dependent variable (its value is determined by what is done with the independent variable) and is used for the vertical axis (*y*-axis). Since the mentioned positions may vary, labeling is important.

3 Give each student a sheet of graph paper. Show a transparency of the graph paper. Have students recreate the graph below with you. Explain how to draw a horizontal and vertical axis. Show them how to title the graph and how to label and number the axes. Explain that the numbering can be on every square, every other square, or whatever seems convenient for the information you are trying to convey. Tell students that the way the graph is numbered is called the *scale* of the graph. The numbering on each axis must increase by the same amount for each square. The horizontal and vertical axes do not need to be numbered using the same scale.

4 Show students how to move to the desired number of CDs on the horizontal axis and then up to the correct amount of money on the vertical axes. Have them each make a dot on that spot.

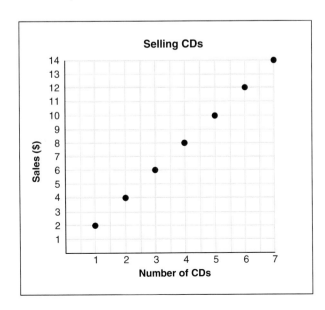

© Shell Education #50616—30 Mathematics Lessons Using the TI-15 113

A New Look (cont.)

Part One
Explaining the Concept (cont.)

5 Ask students: *Is this graph a good way to display the data? What are some advantages of using a graph over using a table?* Guide students toward the idea that the graph can be an easier and clearer way to gain a quick understanding of the data.

Applying the Concept

1 Distribute copies of the *Dozens and Dozens* activity sheet (pages 117–118; page117.pdf). Remind students how to use [Opl] to set up the repeated multiplication to find the amount of money Serena will make on her cookies. They will need to press [Opl] [×] [.] [5] [0] [Opl]. Tell them that they will need to enter the number of cookies and then press [Opl].

2 Have students discuss the organization of the graph—labels, axes, scale, etc. Ask them how to mark a point for 1 cookie and $0.50. They should move to line 1 on the horizontal axis and then up halfway between 0 and 1 on the vertical axis.

3 After students have completed the graph and questions a–d, discuss how the amount of money earned relates to the number of cookies sold. (*For each cookie sold, $.50 is earned.*) Ask students: *How is this represented on the graph?* (*The amount of money earned on the y-axis increases by $.50 for each cookie sold, as represented by the x-axis.*) Students should understand that the points are on a line because the graph goes up by the same amount for each cookie.

4 Make sure that students know how to fill the values in for the second table using the [Opl] key. After they have completed the table, have students discuss the organization of the graph—labels, axes, scale, etc. Give time for completion of the second graph.

5 When students have completed their graphs, discuss the meaning of the point where the number of the cookies and the amount of money are both zero. Then ask: *How many cookies are sold when the earnings are $45?* (9 dozen)

Part Two
Explaining the Concept

1 Distribute copies of *The Cookie Business* activity sheet (pages 119–121; page119.pdf). Explain to students that they are going to investigate expense, income, and profit in a business.

2 After reading and discussing the problem, help students set up the computation for Donna's expenses using [Opl] and [Op2]. Have students press [Opl] [×] [2] [Opl]. This will compute the cost of ingredients. Then have students press [Op2] [+] [3] [0] [0] [Op2]. This will add on the monthly expenses. Explain that when they enter the number of dozens of cookies and then press [Opl] and [Op2], the TI-15 will multiply the number by 2 and then add 300 automatically.

A New Look (cont.)

Part Two
Explaining the Concept (cont.)

3 Before students fill in the chart, ask them to enter a number and press [Op2] and then [Op1] (reverse the order) to see what happens. Make sure they understand that if they press these buttons out of order, the TI-15 will add first and then multiply, which will result in a completely different answer. Have each student explain to a partner the problem in their own words and why they are multiplying by 2 and adding by 300.

4 Have students discuss the organization of the graph—labels, axes, scale, etc. Ask them how they should move on the horizontal axis to find 50 dozen. Point out that although the numbers are increasing by hundreds, each tick mark represents 50. After students complete the graph and table in problem 1, have them turn to partners and summarize in their own words what is represented in the graph and table and how the values in each month change. Then have students answer questions 1a–1c on page 120.

5 Before completing problem 2 (page 120), discuss the difference between *income* and *profit*. Next, discuss the questions for problem 2. Prepare them for problem 2b, by helping them find Donna's profit for 500 dozen cookies. Then, use [Op1] and [Op2] to complete the table on page 121 as a class. Have students represent the profits in the graph. Based on the needs of your students, you may choose to have them use the grid provided or you may wish to have them create a graph on a full sheet of graph paper. This would allow for the vertical axis to be marked for every $50 or $100.

Applying the Concept

1 Discuss the graph. Have students draw a line that passes through the plotted points. Have them use the graph to determine whether there is a profit or loss when 500 cookies are sold. (*profit*) Ask them how the graph helps them see this. (*The graph is above the horizontal axis.*) Have them estimate the profit and then confirm it on the TI-15. ($950)

2 After students have completed the questions about the graph, discuss the answers as a class. The remaining problems on *The Cookie Business* activity sheet will be completed in Part Three.

Part Three
Explaining the Concept

1 Have students answer the questions regarding the *Donna's Profits* graph (page 121) independently. Walk around the room to assist students as needed.

2 Review the answers as a class. Conclude with a question such as: *How many cookies are needed to make a profit of $1,200?* (600)

Power Lesson 2

A New Look (cont.)

Applying the Concept

1. Have students work in pairs or small groups to invent businesses. Students should research costs on the Internet. Ask them to determine their monthly expenses, the cost to make items, and the selling price for those items.

2. Distribute graph paper. Have them make graphs for their businesses. Students should then trade graphs and determine another group's monthly expenses and the profit per item. Have a few students put their graphs on transparencies and explain them to the class.

Differentiation

- **Below Grade Level—** Allow students to use fake money to simulate the real-world scenarios. This will help them to accurately complete the charts.

- **Above Grade Level—** Have students write about what they think might happen if the profit or cost changed at any point during Donna's business scenario.

Extending the Concept

- Explain to students that points can be identified as ordered pairs [e.g., (5, 2)], where the first number is the distance on the horizontal axis and the second is the distance on the vertical axis. Tell them that each number is called a *coordinate*. Ask them to express points from the lesson as ordered pairs.

- Show students how to write formulas for the data in the tables. Show them that they can use the formulas to write equations for the lines.

Name _____

Dozens and Dozens

Directions: Read the text carefully and complete the graph as instructed.

Mrs. Acousta's class is making cookies and brownies for the school bake sale. Mrs. Acousta wants students to provide information that shows how much money they will make.

1. Serena is in charge of oatmeal cookies. She wants to charge $0.50 per cookie. Use your TI-15 and follow the steps below to complete the chart to see how much money the class would make on oatmeal cookies.

   Step 1   Press [Opl] [×] [.] [5] [0].

   Step 2   Press [Opl].

   Step 3   Enter the number of cookies.

   Step 4   Press [Opl].

   Step 5   Continue to enter the number of cookies and press [Opl].

Number of Cookies	1	2	3	4	5	6	7	8	9
Earnings in Dollars									

Mrs. Acousta wants a more visual display and asks Serena to make a graph. She tells Serena to have the horizontal axis be the number of cookies and the vertical axis be the earnings. Use the grid below to make Serena's graph.

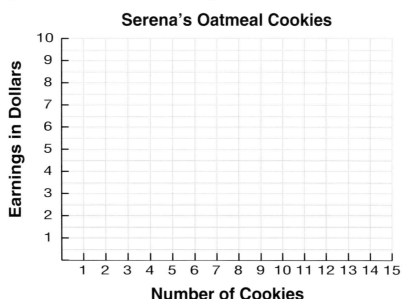

Serena's Oatmeal Cookies

© Shell Education   #50616—30 Mathematics Lessons Using the TI-15   **117**

Dozens and Dozens (cont.)

Directions: Use Serena's graph on the previous page to answer the questions below.

a. For every cookie sold, how much do Mrs. Acousta's earnings increase?

_____

b. Serena sees that all of the points lie on a line. Why does that make sense?

_____

c. Use a pencil to draw in the line very lightly. Put a point on the line where the number of cookies would equal 13. What is the cost of 13 cookies?

_____

d. Suppose the number of cookies is 0. What is the number of dollars?

_____

2. Fritz is in charge of chocolate chip cookies. He decides to sell cookies only by the dozen. He will charge $5.00 per dozen. Use [Op1] key to complete the chart below.

Dozens of Cookies	1	2	3	4	5	6
Earnings						

Fritz also needs to create a graph. Give the graph a title. Label the axes. Determine a scale for the *x*-axis and the *y*-axis. Then make Fritz's graph.

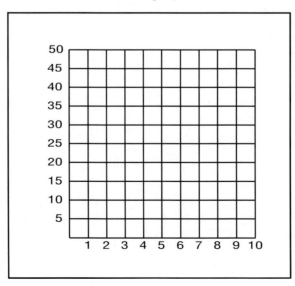

Name _____

The Cookie Business

Directions: Read the text, complete the graphs as instructed, and answer the questions.

1. Donna is starting a business making and selling cookies. She is going to run the business out of her home, so her costs will be fairly low. She estimates that her monthly expenses will be about $300. The cost of the ingredients for each dozen cookies will be $2.00. As Donna's business becomes established, she hopes to increase the amount of cookies she sells by 50 dozen each month. Use your TI-15 and follow the steps below to calculate the total amount of expenses for each month.

   Step 1 Press [Op1] [×] [2].
   Step 2 Press [Op1].
   Step 3 Press [Op2] [+] [3] [0] [0].
   Step 4 Press [Op2].
   Step 5 Enter the number of dozens of cookies.
   Step 6 Press [Op1] [Op2].

Repeat steps 5 and 6 to complete the table below to compute Donna's possible expenses for each month. Then use the grid below to graph Donna's expenses.

Month	Jun.	Jul.	Aug.	Sept.	Oct.	Nov.	Dec.	Jan.	Feb.
Dozens of Cookies	50	100	150	200	250	300	350	400	450
Expenses in Dollars									

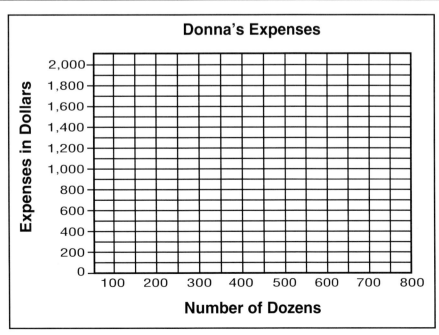

© Shell Education

The Cookie Business (cont.)

Directions: Use the *Donna's Expenses* graph on the previous page to answer the questions below.

a. How much do the expenses go up each time the number of dozen of cookies increases by 50?

_____

b. What would Donna's expenses be if she made 700 dozen cookies?

_____

c. What does it mean when the line crosses the vertical axis?

_____

2. Donna is trying to decide how many dozen cookies she needs to sell to make a profit.

   a. If Donna sells her cookies for $4.50 per dozen, how much does she make per dozen if she subtracts out the $2.00 per dozen for the ingredients?

   _____

   b. Donna remembers that she has the other costs that total $300 per month. What would her profit be if she sold 500 dozen cookies? How did you find it?

   _____

   _____

   _____

   c. How can you use [Op1] and [Op2] to find the profit for any number of dozen of cookies?

   _____

3. Use [Op1] and [Op2] to calculate Donna's profit each month. Complete the table on the next page as you make each calculation.

The Cookie Business (cont.)

Directions: Use your TI-15 and the [Op1] and [Op2] keys to complete the chart below.

Months	Jun.	Jul.	Aug.	Sept.	Oct.	Nov.	Dec.	Jan.	Feb.
Dozens of Cookies	50	100	150	200	250	300	350	400	450
Profit									

Directions: Graph the data on the grid below starting with 50 dozen.

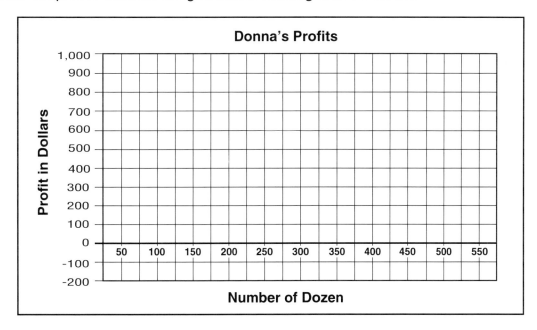

Directions: Use the *Donna's Profits* graph to answer the questions below.

a. What is happening to Donna's business if she sells only 50 dozen cookies?

_____

b. What happens between 100 and 150 dozen?

_____

c. What is Donna's profit or loss if she does not sell any cookies for a month?

_____

d. Estimate the number of dozen of cookies where Donna "breaks even."

_____

© Shell Education

Lesson 11

Mini Me

Overview

Students will use measuring tools and calculators to make stick-figure scale drawings of themselves.

Mathematics Objective

Understands how scales in maps and drawings show relative size and distance.

TI-10 Functions

- Fractions
- Fractions to decimals
- Operations

Materials

- TI-15s
- rulers
- string
- tape measures or yardsticks
- blank paper
- *Scaled Down* activity sheet (pages 125–126; page125.pdf)
- *Scaled Up* activity sheet (page 127; page127.pdf)

ABC Vocabulary

Complete the *Chart and Match* (page 19) vocabulary activity using the words below. Definitions for these words are included on the Teacher Resource CD (glossary.pdf).

- scale
- scale drawing

Warm-up Activity

1. Before class, find your height to the nearest inch. Then, calculate one-eighth of your height (e.g., If you are 5' 8" tall, then one-eighth your height is $8\frac{1}{2}$ inches).

2. Tell students that in a photograph, your height measures the number you found above (e.g., $8\frac{1}{2}$ inches).

3. Ask them to guess how many times taller than that you really are.

4. Have students find heights that correspond to their guesses using their TI-15s. Show them how to enter your height into their TI-15s by entering the whole number, pressing [Unit], entering the numerator, pressing the [n] key, entering the denominator, and then pressing the [d] key. After the mixed number is entered, they can multiply by the scale factor (e.g., If students think you are 10 times taller than your photograph, they would multiply $8\frac{1}{2}$ by 10 to get your "real" height of 85 inches).

5. Ask students what their guesses are. Have a student measure your height. Have students who were correct hold up their TI-15s.

#50616—30 Mathematics Lessons Using the TI-15 © Shell Education

Mini Me (cont.)

Explaining the Concept

1 Ask students if they have ever seen a scale drawing. Discuss why scale drawings are used. Ask them how the shape of the scale drawing compares to the shape of the original object.

2 Tell students that they are going to make stick-figure scale drawings of themselves.

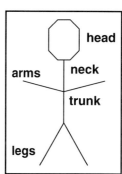

3 Have them work with partners to measure the body parts for the stick figures. They should follow these instructions for measuring: Measure the head from the top of the head to the bottom of the chin. For the upper line segment (neck), measure from the bottom of the head along the spine to the shoulders. For the trunk, measure from the underarm to the top of the leg along the side of the body. Measure the arms from the top of the shoulder to the fingertips. For the legs, measure from the top of the leg to the floor. Tell students to measure to the nearest quarter inch. Show students how to use string to find a length, and then use a yardstick or tape measure to find the measurement. The string is easier to use than a tape measure.

4 Ask students how they can make sure that the scale drawings have the correct measurements. They should know that they will divide or multiply each measurement by the same number.

5 Work through an example with students. Tell them that a student's arm measures $16\frac{1}{2}$ inches. Ask them how to find its length on a drawing if the drawing is going to be one-fourth the size of the original. Show them that they can multiply by $\frac{1}{4}$ or divide by four.

6 Have them use their TI-15s to find the length on the drawing using multiplication and division. Be certain that they know how to enter the fractions by using the keystrokes [1] [6] [Unit] [1] [n] [2] [d] [×] [1] [n] [4] [d] [Enter]. It is not necessary to use the [d] key after the denominators, but it may emphasize the structure of the fraction for students. (Multiplying by 0.25 would cause the answer to be displayed as a decimal, making it difficult to use inches for the drawing. The result will be $4\frac{1}{8}$ inches.)

7 Distribute copies of the *Scaled Down* activity sheet (pages 125–126; page125.pdf). Give students time to complete it. When students have finished, have them lightly write their names on the backs of their sheets.

Mini Me (cont.)

Applying the Concept

1. Choose one drawing from the *Scaled Down* activity sheet. Ask students how they could find the real measurements of the student shown. Remind them that multiplying by 4 "undoes" dividing by 4 or multiplying by $\frac{1}{4}$.

2. Have students trade stick-figure drawings with partners and measure the parts of the drawings they receive. Do not have them trade sheets with the classmates who were their partners in the measuring section of the activity. Using their *Scaled Up* activity sheets (page 127; page127.pdf), have students find the actual measurements of the students on their sheets.

Differentiation

- **Below Grade Level—** Choose some numbers, and have students multiply by one-fourth and divide by 4 to emphasize that these two operations are equivalent.

- **Above Grade Level—** Have students look at a map or a blueprint where a scale is given in terms of different units of measure (e.g., 1 inch = 5 miles).

Extending the Concept

Have students bring in three-dimensional models, such as model cars or anatomical models of dinosaurs. Have the students find the actual dimensions.

Name _____

Scaled Down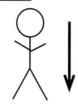

Directions: Follow the steps below to make a stick figure drawing based on the lengths of parts of your body. Draw your figure on a separate sheet of paper.

Your figure should look something like this:

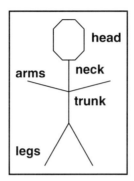

Your drawing is going to be one-fourth the size of you. Find the length of each part of your drawing by following either set of instructions. Try using both sets of instructions. Place the measurements for your drawing on the chart on the next page.

Multiplication	Division
1. Enter the whole number.	1. Enter the whole number.
2. Press [Unit].	2. Press [Unit].
3. Enter the numerator.	3. Enter the numerator.
4. Press [n].	4. Press [n].
5. Enter the denominator.	5. Enter the denominator.
6. Press [d].	6. Press [d].
7. Press [x].	7. Press [÷].
8. Press [1] [n] [4] [d] [Enter].	8. Press [4] [Enter].

© Shell Education #50616—30 Mathematics Lessons Using the TI-15

Scaled Down (cont.)

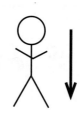

Part of Drawing	Measurement (to the nearest $\frac{1}{4}$ inch)	Length on Scale Drawing (to the nearest $\frac{1}{16}$ inch)
head width		
head length		
neck		
trunk		
arms		
legs		

1. Why can you use either set of instructions and get the same answer?

    _____

2. Suppose you wanted the drawing to be $\frac{1}{8}$ the size of you. What would be the last steps on the multiplication and division instructions?

    _____

3. Why can't you find the total height of your drawing from the measurements?

    _____

4. What numbers should you add together that will give you the largest possible height for your drawing?

    _____

5. What is the largest possible height of your drawing?

    _____

Name _____

Scaled Up

Directions: In the activity sheet *Scaled Down*, you found the measurements to make a scale drawing of a stick figure of yourself from your body measurements.

Trade stick-figure drawings with a partner and measure the parts of the drawing. Do not trade sheets with the person who was your partner in the measuring section of the *Scaled Down* activity.

Remember that you multiplied by $\frac{1}{4}$ or divided by 4 to find the lengths for the stick figure. Find the actual measurements of your partner using his or her stick drawing. Complete the chart below. Then see if these measurements are close to correct.

Part of Drawing	Length on Scale Drawing (to the nearest $\frac{1}{16}$ inch)	Measurement (in inches)
head width		
head length		
neck		
trunk		
arms		
legs		

1. How did you find the measurements of the person?

   _____

2. Were the measurements the same? _____

3. Why might the answers be close but not exactly the same?

   _____

Lesson 12

On the Ball

Overview

Students will use linear measurement and calculators to investigate proportionality and determine the constant ratio between similar figures.

Mathematics Objective

Uses, measures, and creates scales for scale drawings.

TI-15 Functions

- Decimals to fractions
- Fractions
- Operations

Materials

- TI-15s
- two pictures; one an enlargement of the other
- centimeter grid paper
- centimeter rulers
- *Big and Small* activity sheet (pages 131–132; page131.pdf)
- *Another Look* activity sheet (page 133; page133.pdf)

 Vocabulary

Complete the *Which Statement Is Accurate?* (page 17) vocabulary activity using the words below. Definitions for these words are included on the Teacher Resource CD (glossary.pdf).

- decimal
- enlargement
- fraction
- proportion
- ratio
- reciprocal
- reduction
- similar

 Warm-up Activity

1. Show students two pictures, one an enlargement of the other. Ask students to estimate the number by which they could multiply the measurements of the smaller picture to find the corresponding measure on the larger picture.

2. Use centimeters to measure the length and width of the smaller picture. With their TI-15s, have students use the numbers they chose in step 1 and multiply them by the measured length and width. This will provide students with their predicted dimensions of the larger picture.

3. Measure the actual length and width of the larger picture. Have students evaluate their predictions to see how close they were to the actual dimensions of the picture.

On the Ball (cont.)

Explaining the Concept

1 Trace the triangles below and show them to the students on the overhead projector. The second is an enlargement of the first. The base of the third triangle is the same length as the base of the second.

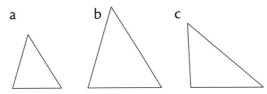

2 Ask students: *Which figure is an enlargement of the first?* If students guess correctly, ask them: *How do you know triangle c is not an enlargement of triangle a?* Students may have difficulty explaining how they can tell which is an enlargement. Explain that the second figure is proportional to the first. This means that you could multiply the length of each side of the first by the same number to get the length of the side on the second.

3 To the nearest millimeter, measure the lengths of one pair of corresponding sides of triangle a to triangle b. Show students the ratio of the side lengths of triangle a and triangle b in the form a/b. Have students use their TI-15s to enter the measurement from triangle a, press [n], enter the measurement from triangle b, press [d], and press [Enter] to see the fraction. Then have them press [Fix] [0.01] [F↔D] [Enter] to see the decimal form to the nearest hundredth. Write the results on the board or overhead. Also, using the decimal form, show the ratio in the form a:b (as in 1:1.54). Then repeat the process using b/a instead. Ask: *Which ratio would you use to find a side from triangle b if you knew the corresponding side from triangle a?* (b/a). Have students confirm this with their TI-15s with both the fraction and the decimal forms of the ratio.

4 Measure a different side on triangle a. Ask: *Which side on triangle b corresponds to it? How can we predict the measurement on triangle b?* Have them multiply this second side of triangle a with the ratio b/a used previously. Have them measure the corresponding side of triangle b and compare their findings. Ask: *Why are the values close but not exactly the same?* Explain that measuring with a ruler is imprecise, so the results will not be perfect.

5 Distribute copies of the *Big and Small* activity sheet (pages 131–132; page131.pdf) to students. Help them determine the first ratio in fraction form. Remind them that they may change to a fraction by pressing [F↔D]. When students have finished the activity sheet, discuss their findings. Ask them how they might agree on the correct ratio.

Lesson 12

On the Ball (cont.)

Applying the Concept

1. Distribute copies of the *Another Look* activity sheet (page 133; page133.pdf) to students.

2. When students have finished, discuss the results. Have them write a summary of what they have learned. Tell them to include details such as how one can predict that a ratio is greater than one, that a ratio is less than one, and that a ratio equals one.

Differentiation

- **Below Grade Level—** Have the students begin by looking at rectangular figures on grid paper. Ask them to double the number of squares on each side and figure the ratio. Then, have them triple the number of squares. Repeat with a rectangle that is 6 x 12. Have them simplify the size of the rectangle by $\frac{1}{2}$ and $\frac{1}{3}$.

- **Above Grade Level—** Show students how to write a proportion to find the missing side of two similar figures. Give them numbers where it is easy to see the relationships.

Extending the Concept

Have students explore the relationship between the ratios that compare a smaller object to an enlargement of the object with the ratio that you would get if you compared a larger object to a reduction of the object. Discuss why these ratios are reciprocals.

Name _____

Big and Small

Directions: Answer the questions below.

Figure A

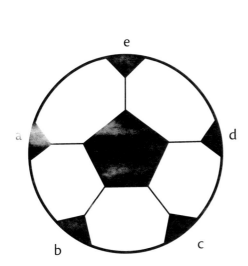

Figure B

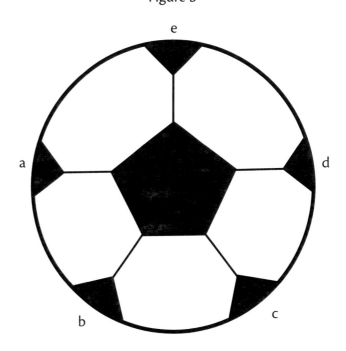

1. Do you think the figures are proportional? _____

2. If the balls are proportional, would the ratio of Figure A to Figure B be larger or smaller than 1?

   _____

3. If proportional, would the ratio of Figure B to Figure A be larger or smaller than 1?

   _____

4. Predict which number you should multiply a measurement on Figure A by to find the length of the corresponding measure on Figure B.

   _____

© Shell Education   #50616—30 Mathematics Lessons Using the TI-15   **131**

Lesson 12

Big and Small (cont.)

Directions: Choose four parts on the two figures that you think are corresponding parts. Measure them and record the locations and the measurements in the first three columns in the chart below. Then answer the questions.

Part Measured	Measurement in Figure A	Measurement in Figure B	Ratio in Fraction Form	Ratio in Decimal Form

5. Find the ratio of the measurement for the first part of Figure A to the measurement for the first part on Figure B. Record it in the fourth column.

6. Change the ratio from a fraction to a decimal using the [F↔D] key. Record the fraction in the last column.

7. Repeat steps 1 and 2 for the rest of the measurements.

8. Did you get the same ratio, decimal form or ratio form, for every measurement? Why or why not?

_____

9. If a measurement on Figure A is 20 mm, what is the corresponding measurement on Figure B?

_____

10. If a measurement on Figure B is 20 mm, what is the corresponding measurement on Figure A?

_____

132 #50616—30 Mathematics Lessons Using the TI-15 © Shell Education

Name _____

Another Look

Directions: Answer the questions.

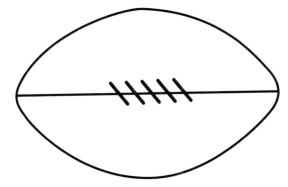
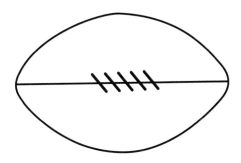

1. Is the football on the right an enlargement or a reduction of the one on the left?

   _____

2. Guess the ratio you will get when you compare the first football to the second football.

   _____

3. Choose four corresponding parts on the two similar figures. Measure them and record them on the chart below. Complete the ratio columns comparing the 1st picture to the 2nd picture.

Part Measured	Measurement in 1st picture	Measurement in 2nd picture	Ratio in Fraction Form	Ratio in Decimal Form

4. Find the average of the ratios for both the decimal and fractional values.

   _____

   _____

Lesson 13

It's on the Map

Overview

Students will learn to estimate distances on maps and to draw maps.

Mathematics Objective

Understands how scales in maps and drawings show relative size and distance.

TI-15 Functions

- Memory
- Operations

Materials

- TI-15s
- grid paper with four or five squares per inch
- rulers
- map of your local area
- *Math Country Map* (mathtown.pdf)
- *Mapping Measurement Mountain* activity sheet (pages 137–138; page137.pdf)
- *My Math Map* activity sheet (page 139; page139.pdf)

Vocabulary

Complete the *Total Physical Response (TPR)* (page 18) vocabulary activity using the words below. Definitions for these words are included on the Teacher Resource CD (glossary.pdf).

- compass
- east
- north
- scale
- south
- west

Warm-up Activity

1. Tell students that two cities are shown on a map as 3.5 inches apart.

2. Ask them if they know how many miles apart the cities are.

3. Ask them what they need to know to determine the distance between the two cities.

4. Ask a student to create a scale for the map (e.g., the number of miles are represented by 1 inch).

5. Have students use that scale and their TI-15s to find the actual distance between the two cities. Students will need to use multiplication to calculate the actual distance.

It's on the Map (cont.)

 ## Explaining the Concept

1 Ask students if they have ever used maps to plan trips. Ask them what kind of information they can get from maps. Pass around a map of your local area. Ask volunteers to point out various features and locations on the map. They may point out such things as highways, roads, airports, cities, rivers, lakes, a compass, or scale.

2 Project the map from page 138 on the overhead or document projector. (If using an overhead projector, a transparency of the map should be created prior to teaching the lesson.) Ask students to identify the northernmost city shown on the map. *(Kilogram Valley)* Then, ask them to find the cities that are farthest apart east, west, and south. *(Ouncesville; Tonsville; Poundsville)*

3 Ask them to plan at least two routes that go from Tonsville to Abacus Town. Ask if they can find alternate routes.

4 Have students look at the scale on their maps. Ask them if they can tell at a glance how many miles are represented by one inch. Discuss how you might use the map to find the number of miles in each of the routes from Tonsville to Abacus Town. Students might suggest marking distances on pieces of paper and comparing them with the scale.

5 If no one suggests finding the number of miles per inch, suggest that they measure the number of inches that represents 50 miles on the scale, and then find the number of miles per inch. This should be about 2 inches. Show students how to divide 50 by 2 by pressing [5] [0] [÷] [2] [Enter]. Show them how to store the number in the memory by pressing [▶M], and then [Enter]. Show how to recall the number in a problem such as 3 x 25. Press [3] [×] [MR/MC] [Enter]. Explain that placing the number in the memory saves a little time and prevents errors in entering the numbers.

6 Distribute copies of the *Mapping Measurement Mountain* activity sheet (pages 137–138; page137.pdf) to students. Tell the students that they are going to plan a tour of the Measurement Mountain region. Have the students work on their activity sheets in pairs.

7 Ask for volunteers to put their routes and distances on the board. Ask if anyone has a total distance more than those shown. Ask if anyone has a shorter total distance. Put those on the board, as well. Ask if anyone has the same route but a different distance. Ask students to account for the differences.

Lesson 13

It's on the Map (cont.)

Applying the Concept

1. Distribute copies of the *My Math Map* activity sheet (page 139; page139.pdf) and pieces of grid paper to students.

2. Have students work in groups of two or three. Tell students that they need to make scales for the maps and place the cities on them according to the directions on the activity sheet.

3. As an example, give students a scale of 1 inch = 5 miles. Ask them how to represent a distance of 47 miles. Warn them to be sure to consider the distances between cities before choosing a scale.

4. Draw the compass points on the board to remind students of the directions. Remind them to indicate which way is north on their maps.

Differentiation

- **Below Grade Level—** Have students use straight-line distances between the cities. Have the class compare the straight-line distances with the distances along the road. Discuss with the class the meaning of "as the crow flies."

- **Above Grade Level—** Have students create maps and give directions using words and symbols such as SE, NW, etc.

Extending the Concept

Have students plan real trips in an area near to their homes. Have them investigate local speed limits and estimate the amount of time it would take to complete the trips.

#50616—30 Mathematics Lessons Using the TI-15 © Shell Education

Name _____

Mapping Measurement Mountain

Directions: Use the map on the next page to plan a trip to the Measurement Mountains region. You want to begin in Kilogram Valley and visit Rulerville, Compassburg, Ounceville, Calculator Town, and Abacus Town. You do not need to visit them in that order.

1. Plan a route beginning and ending in Kilogram Valley that visits each of the cities listed above. On the chart below, show the order in which you will visit each city and the highways that you will use. You may use more than one road between towns.

City	Road	Inches	Miles	Kilometers
Kilogram Valley				
		Total		

2. How many miles are represented by each inch? Place that number in the memory of the TI-15 by entering the number and then pressing [Enter] [▶M] [Enter]. Use it to find the number of miles traveled on each leg of the trip by entering the measurement and pressing [×] [MR/MC] [Enter].

3. Find the total number of miles traveled on your trip. _____

4. How many kilometers are represented by 1 inch? Store that number in the memory of the TI-15 by entering the number and then pressing [Enter] [▶M] [Enter]. Use it to find the number of miles traveled on each leg of the trip by entering the measurement and pressing [×] [MR/MC] [Enter].

5. Find the total number of kilometers traveled. _____

Lesson 13

Mapping Measurement Mountain (cont.)

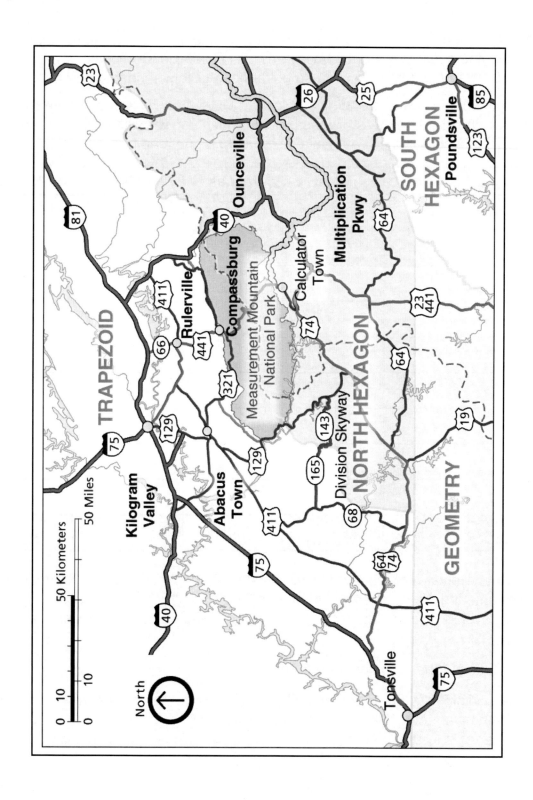

Name _____

My Math Map

Lesson 13

Directions: Use the information to map the towns on a piece of graph paper. Find the number of squares per inch on the paper to help you determine the scale you wish to use.

- Addend is the northernmost town on the map.
- Numeral is directly south of Addend and is 50 miles from Addend.
- Fractionville is directly east of Numeral and is 18 miles from Numeral.
- Semicircle is directly south of Fractionville and is 50 miles from Fractionville.
- Radius Ranch is directly west of Semicircle and is 25 miles from Semicircle.
- Triangleburg is directly west of Radius Ranch and is 62 miles from Radius Ranch.
- Highway 3.14 begins at Addend and connects all of the towns on this map.

1. What scale are you going to use on your map? _____

2. Place the towns on your map. Be sure to indicate the scale and which direction is north on your map.

3. Use your scale to find the distance on the map between each of the towns while traveling on Highway 3.14.

   a. Addend to Fractionville _____

   b. Numeral to Radius Ranch _____

4. Highway 2^2 is being built to connect Addend to Triangleburg. Use your map to estimate the distance the new highway will cover.

   _____

5. Highway 111 connects Fractionville and Radius Ranch. Use your map to estimate the distance Highway 111 will cover.

   _____

6. If you want to travel from Addend to Triangleburg, is it faster to take Highway 3.14 or the new Highway 2^2? Why or why not?

   _____

© Shell Education #50616—30 Mathematics Lessons Using the TI-15

Lesson 14

Scale That Structure

Overview

Students will use proportions to create scale drawings of structures.

Mathematics Objective

Understands how scales in maps and drawings show relative size and distance.

TI-15 Functions

- Constant feature
- Operations

Materials

- TI-15s
- butcher paper (36" wide)
- string and rulers
- picture of Taj Mahal (mahal.jpg)
- picture of White House (house.jpg)
- picture of Arc de Triomphe (arc.jpg)
- picture of Buckingham Palace (palace.jpg)
- *Around the World* activity sheet (page 143; page143.pdf)
- *My Structure* activity sheet (pages 144–145; page144.pdf)

 Vocabulary

Complete the *Sentence Frames* (page 16) vocabulary activity using the words below. Definitions for these words are included on the Teacher Resource CD (glossary.pdf).

- dimensions
- proportional
- scale

 Warm-up Activity

1. Tell students that a local community group wants to make a statue of the mayor. The mayor is 75 inches tall, but the statue is going to be 15 feet tall.

2. Ask students what number of inches they think will be represented by 1 foot on the statue. Have students use their TI-15s to divide 75 by 15.

3. Explain to students that the scale is 1 foot = 5 inches.

4. Tell students that the statue of the mayor is holding a book that is 2 feet long. Ask them to find the length of the actual book.

5. Review the following TI-15 keystrokes with students. Now that they know the actual book is 10 inches, they can practice by entering the multiplication of the scale ratio. Have them clear the memory by pressing (On) and (Clear) at the same time. After pressing (Clear) again, have students press [OpI] [×] [5] [OpI]. By pressing [2] [OpI], 2 × 5 for a product of 10.

#50616—30 Mathematics Lessons Using the TI-15 © Shell Education

Scale That Structure (cont.)

Explaining the Concept

1 Show students pictures of the following structures: the Taj Mahal in India, the White House in Washington D.C., the Arc de Triomphe in France, and Buckingham Palace in England. Tell students a little bit about each structure as you show them the pictures. Then tell students that they are going to use a process similar to the one in the warm-up activity to make scale drawings of these structures.

2 Distribute copies of the *Around the World* activity sheet (page 143; page143.pdf). Go over the description of the project with students. Tell the class that they are going to be put into one of four groups and that each group will be responsible for creating one scale drawing. This scale drawing will serve as the backdrop for a play involving different locations from around the world.

3 Ask students why it would be a good idea if they all used the same scale on their drawings. Tell students they will be given a scale to use in order to find the height and length of their structures. Each group will need to change the scale to a decimal.

4 Students will use the given scale to find the dimensions for each of the four structures. Remind students how to find the new heights using their TI-15s.

5 Tell students that they will use butcher paper to create their scale drawing. Explain to students that the width of the paper is 36 inches, so each group will need to decide how long their paper will need to be.

6 Assign each group its own structure. Have each group find the measurements for its scale drawings and make a scale drawing for its structure. When students have finished their drawings, have them switch with another group to make sure their scale is correct.

7 When it has been determined that the drawings for each group's structure are correct, have students look at pictures of their structure in books or online. Have them add details and color to their drawings. You may wish for your students to determine a scale between their drawing and the pictures. With such a scale, they can include more measurements in the drawings beyond just the general height and length.

8 Bring the class back together. Ask students: *If each group sets its drawing side by side, how long will the backdrop be in inches? In feet?* Have students use their TI-15s and work together as a class to find the answer. Have students use the measurements to find a wall on which to post the scale drawings in the school.

Scale That Structure (cont.)

Applying the Concept

1. Distribute copies of the *My Structure* activity sheet (pages 144–145; page145.pdf) and pieces of paper to students.

2. Tell students they will now create their own unique structures. They will decide on the height and length of the structure. They will also get to determine the scale they use. The trick is that they must draw their structure in the 8 in. x 6.5 in. box provided on the activity sheet.

Differentiation

- **Below Grade Level—** Help students write the steps to create a structure. Help them design a template for the sketch using the correct proportions.

- **Above Grade Level—** Have students use clay or modeling compound to create a three-dimensional mini-version of their unique structure.

Extending the Concept

Have students investigate proportions common to the human anatomy, such as the relationship between arm span and height, and then look at examples of famous statues to see if these ratios are used.

Name _____

Around the World

Lesson 14

Directions: Read the information below and complete the activity with your assigned group.

Your class has been chosen to make the backdrop for a play. The play is called *Around the World*. In keeping with the worldly theme, your teacher decided to make the backdrop showing four well-known structures from around the world. To be realistic, the figures on the backdrop must be proportional to the structures themselves.

The actual measurements of the four structures are shown in the table below.

	Taj Mahal	The White House	Arc de Triomphe	Buckingham Palace
Height	213 feet	70 feet	165 feet	79 feet
Length	186 feet	168 feet	148 feet	354 feet

Use the scale 1 inch = 5 feet to keep the structures in proportion with one another. Use the TI-15 to find the height and width of each of the structures. Express the new dimensions as decimals. Round the dimensions to the tenths place. Fill in the chart below.

	Taj Mahal	The White House	Arc de Triomphe	Buckingham Palace
Height				
Length				

Compare your group's chart with the rest of the groups in your class. Make a scale drawing for your group's structure on butcher paper. The butcher paper is 36 inches wide, so you may need to double the paper so your structure will fit.

How long will your butcher paper need to be? _____

Have another group check the measurements of your drawing. If the scale is correct, add detail and color to the scaled drawing of your structure.

© Shell Education #50616—30 Mathematics Lessons Using the TI-15

My Structure

Directions: Follow the steps below to create your own unique structure drawn to scale.

You have been chosen to design a new structure for any country in the world. You will be drawing this structure to scale. The structure can be as tall and long as you choose. You will list the dimensions in feet. You may choose the scale. However, the drawing must fit in the 8 inch by 6.5 inch square on the second page of this activity sheet. Choose your dimensions and your scale wisely!

Step 1 Choose the real-life dimensions of your structure. Remember, the larger the structure, the smaller the scale will have to be. Add your dimensions to the chart below.

Step 2 Choose your scale. You may have to try a few different scales before you find one that will correctly fit on the page. You do not want your drawing to be too small, but it does need to fit on the page. Remember to use your TI-15 to help you!

Step 3 Add your dimensions of the scale drawing to the chart below. Use your TI-15 to find these dimensions.

Step 4 Sketch your drawing.

Step 5 Switch drawings with a partner to see if he or she can find the real-life dimensions of your structure.

Step 6 Answer the questions below. Then add details and color to your drawing.

Scale: _____

	Structure	Scale Drawing
Height		
Length		

My Structure (cont.)

Lesson 15

Painting Flags

Overview

Students will learn the difference between shapes that are similar and congruent and investigate proportions using rectangles on flags.

Mathematics Objective

Understands that shapes can be congruent or similar and that similar shapes can be related proportionally.

TI-15 Function

- Operations

Materials

- TI-15s
- *Flags from Around the World* (flags.jpg)
- *Flag of Greece* (greece.jpg)
- *Similar and Congruent* activity sheet (page 149; page149.pdf)
- *Stripes* activity sheet (pages 150–151; page150.pdf)

 Vocabulary

Complete the *Music Makers* (page 18) vocabulary activity using the words below. Definitions for these words are included on the Teacher Resource CD (glossary.pdf).

- congruent
- dimensions
- equivalent
- foot
- inch
- proportional
- quotient
- ratio
- scale
- similar
- square foot
- square inch

 Warm-up Activity

1. Place students into groups and assign each group one wall.

2. Give each group of students the dimensions of the room's walls. Ask them to use their TI-15s to find the area of their walls.

3. Tell students that a gallon of paint will cover about 400 square feet. Tell them that the paint costs $19.99 per gallon. Ask each group to find the amount of paint needed for its wall and how much it will cost.

4. Come together as a class and figure out how much it will cost to paint the whole room.

Painting Flags (cont.)

Explaining the Concept

1. Ask students to describe the flag of their home countries. Show students the image of the *Flags from Around the World* (flags.jpg) so they can get an idea of the different types of flags.

2. Insert a picture of a flag into a *Microsoft Word*® document. Show students the dimensions of the image. Change the width of the shape. Ask students: *What happens to the height when you change the width?* Change the height of the shape. Ask: *What happens to the width? Why do these different changes occur?*

3. Insert the flag picture twice into the document. Explain the terms *congruent* and *similar*. Make both flags the same length and width. Have students list all the ways the two flags are the same and different. These flags are congruent. Now change the dimensions of one of the flags. Have students list all the ways the flags are the same and different. These flags are similar but not congruent.

4. Discover with students the relationship between proportions (equivalent ratios) and similar shapes. Delete the flags in the document. Have two student volunteers model how to create similar flags by changing the dimensions of the shapes. Compare the dimensions of both flags. With students, find the ratio between length and width for the first flag. Then find the ratio between length and width for the second flag. Compare the ratios. Ask students: *Why are the ratios of length to width the same between both figures?* Students should conclude that the dimensions of similar shapes like flags are proportional, meaning that the ratios are equivalent. Although the dimensions may change, the ratio of length to width remains the same.

5. Ask students: *If the dimensions on the flag are changed, will the dimensions of the stripes change too?* Have students measure length and width of the stripes on two similar flags. Ask: *Do you think the length-to-width ratio of the stripes on the flag will change or stay the same if the dimensions of the flag are changed?*

6. Model how to calculate the area for one of the stripes on the first flag. Ask students: *Is the area of a stripe on the second flag greater than, less than, or equivalent to the areas of the a stripe on the first flag? How will the area of a stripe that is 2 inches wide compare with the area of a stripe that is 1 inch wide?*

7. Distribute copies of the *Similar and Congruent* activity sheet (page 149; page149.pdf) to students. When students have completed the activity sheet, have them explain their mathematical reasoning and understanding of similar and congruent figures and how these relate to the ratio of dimensions of shapes.

© Shell Education

Lesson 15

Painting Flags (cont.)

Applying the Concept

1. Project the image of the Greek flag for students (greece.jpg). Distribute copies of the *Stripes* activity sheet (pages 150–151; page150.pdf) to students.

2. Explain that they are going to investigate the area of the stripes on different sizes of Greek flags. When students have completed the activity sheet, review the work and ask them what they have found out about proportional figures.

3. Help students take a closer look at question 9. Have them draw a square with both the length and the width labeled 20 ft. Have them find the area. (*400 sq. ft.*) Have them draw another square. Ask: *How many inches are in 20 feet?* (*240 inches*) Have them label the length and width. Have them find the area. (*57,600 sq. in.*) Show that this value matches the value they should have found for question 5 when they multiplied 400 and 144 to find the area.

Differentiation

- **Below Grade Level—** Have students use grid paper to draw rectangles with the number of units equal to the measures of the sides of the rectangles in the *Stripes* activity sheet.

- **Above Grade Level—** Ask students to find the volume of a box that is 2 in. x 3 in. x 4 in. Ask them to find the dimensions and the area if all the dimensions increase by 2. Continue this for 3 and 4. Have them investigate this relationship between similar figures and volume.

Extending the Concept

Ask students what they think the relationship between the areas of two flags would be if the length of one flag were 17 times greater than another flag.

Name _____

Similar and Congruent

1. What does it mean for two shapes to be congruent?
   _____

2. What does it mean for two shapes to be similar?
   _____

3. Are the flags below similar or congruent? How do you know?
   _____

   12 units | Flag A | Flag B | 24 units

   22 units

   22 units

4. What are the dimensions of both flags?
   _____

5. What is ratio of length to width for Flag A? for Flag B?
   _____

6. Are the ratios the same or different? Why or why not?
   _____

7. What is the area for both flags?
   _____

8. On a separate sheet of paper, draw and label two rectangles that are both similar to Flag A. What is the ratio of the length to width for the first new rectangle? for the second?
   _____

9. On a separate sheet of paper, draw and label a rectangle that is similar to Flag A but with a width of 132 units. What is the length of the new rectangle?
   _____

Stripes

Directions: Nikolaus is going to be part of a team painting a mural that shows the different flags from different countries. Nikolaus is from Greece and has been put in charge of painting the blue and white stripes on the flag that represents Greece.

The dimensions of the flag Nikolaus is going to paint need to be similar to the actual flag, but not congruent. He needs to create a scale to keep the dimensions proportional to the actual flag. Nikolaus looks up the measurements for the stripes and finds that on the long stripes, the ratio of width to the length is 1 inch to 24.7 inches. For the short stripes, the ratio of the width to the length is 1 inch to 14.82 inches. The widths of all the stripes are the same.

Nikolaus needs to find how much paint he will need. He knows that the amount of paint needed depends upon the area of the stripes. He decides to start with the width of a stripe at 1 inch, find the length of the stripes, and then the area of the stripes. Then, he will change the width of the stripes to see how the area will change. Complete the chart below.

Long Stripes		
Width	**Length**	**Area**
1 in.		
2 in.		
3 in.		
4 in.		
Short Stripes		
Width	**Length**	**Area**
1 in.		
2 in.		
3 in.		
4 in.		

Stripes (cont.)

Directions: Follow the steps below.

1. What is the quotient when you divide the area of a long stripe with a width of 2 inches by the area of a long stripe with a width of 1 inch?

2. What is the quotient when you divide the area of a long stripe with a width of 3 inches by the area of a long stripe with a width of 1 inch?

3. What is the quotient when you divide the area of a long stripe with a width of 4 inches by the area of a long stripe with a width of 1 inch?

4. Repeat numbers 1–3 with the short stripes. Did you get the same results?

5. What would you multiply the area of the 1-inch stripe by to find the area if the stripe were 10 inches wide? _____

Nikolaus went shopping for paint. The paint cans indicate that a gallon of paint covers 400 square feet. He needs to know how many square inches the flag will cover.

6. Draw a square. Label each side 1 foot. What is the area of the square? _____

7. Replace the 1-foot measurements with 12 inches. What is the area of the square?

8. How many square inches are in a square foot? _____

9. How could Nikolaus find out how many square inches a gallon of paint would cover?

Power Lesson 3

Shape Investigations

Overview

Students will investigate two-and three-dimensional shapes. They will determine perimeter, area, surface area, and volume when appropriate.

Mathematics Objective

Predicts and verifies the effects of combining, subdividing, and changing basic shapes.

TI-15 Function

- Memory
- Operations

Materials

- TI-15s
- sheets of paper and chart paper
- digital camera
- newspapers or magazines
- scissors, glue, or tape
- rulers, metersticks, or tape measures
- pictures of shapes to use as models
- similar figures
- *Shape Scavenger Hunt* activity sheet (pages 157–158; page157.pdf)

Materials (cont.)

- *Investigating Shapes* activity sheet (page 159; page159.pdf)
- *Measuring Irregular Shapes* activity sheet (page 160; page160.pdf)
- *Designing a Park* activity sheet (page 161; page161.pdf)

 ## Vocabulary

Complete the *Chart and Match* (page 19) vocabulary activity using the words below. Definitions for these words are included on the Teacher Resource CD (glossary.pdf).

- angles
- area
- congruent
- irregular shapes
- parallel
- perimeter
- perpendicular
- polygons
- similar shapes
- surface area
- three-dimensional shapes
- two-dimensional shapes
- volume

 ## Warm-up Activity

1. Give students one minute to draw as many two-dimensional shapes as possible on sheets of paper. Have students label the names of their shapes.

2. Have students use the TI-15 to find the total number of sides or line segments among the shapes that they drew on their papers.

3. Have students share their totals. Write the totals on the board. Have students identify the total of the fewest number of sides or line segments. Have students identify the total of the greatest number of sides or line segments.

Shape Investigations (cont.)

Part One
Explaining the Concept

1 As a whole class or in small groups, have students brainstorm the names of the two-dimensional and three-dimensional shapes that they know. Create two charts, each with three columns. Use one chart to record the names of two-dimensional shapes. Use the second chart to record three-dimensional shapes. Write the names in the first column, draw a picture of the shape in the second column, and list characteristics of the shape in the third column.

2 In order to identify characteristics of their shapes, students will need to be familiar with the following concepts: two-dimensional, three-dimensional, congruent, similar, parallel and perpendicular sides, types of angles, perimeter, area for two-dimensional shapes, and surface-area and volume for three-dimensional shapes.

3 Review these concepts with students. Develop actions to represent the types of angles and parallel and perpendicular lines. Have students play Simon Says using these actions.

4 For those students whose mystery shape is two-dimensional, show students how to calculate the perimeter and area on the TI-15. *(area of a rectangle = length x width; area of a triangle = $\frac{1}{2}$ base x height; perimeter = sum of the length of all the sides of the shape.)*

5 For those students whose mystery shape is three-dimensional, show them how to use the TI-15 to calculate the volume and surface area. Have students research the formulas for surface area and volume for their specific shape. When it is time to begin the *Investigating Shapes* activity, lead students to choose only mystery shapes that are reasonable for their grade level for finding the volume and surface area. Or, for those who choose shapes with complicated volumes or surface areas, be prepared to allow the students to find the perimeter and area of one face.

6 Model how to complete the *Investigating Shapes* activity sheet (page 159; page159.pdf) with sample shapes from around the classroom.

Applying the Concept

1 Have students use their knowledge of two-dimensional and three-dimensional shapes to complete the *Shape Scavenger Hunt* activity sheet (pages 157–158; page157.pdf). As a class, travel around the school to find shapes to complete the scavenger hunt.

2 For the last item on the scavenger hunt, students must take a picture of a shape that they do not know the name of. After the scavenger hunt, they will investigate these unknown shapes.

Shape Investigations (cont.)

Part One (cont.)
Applying the Concept (cont.)

3 Complete *Investigating Shapes* (page 159; page159.pdf) with students. Note: If students do not find unknown shapes during the scavenger hunt, reverse the activity. Fill out the items on shape investigations for a specific shape. Have students use the characteristics listed for that shape to determine the name of that shape. Students then must draw the shape, list three other characteristics for the shape, and provide at least three examples of where they can find that shape in the real world.

4 Students should complete the "Mystery Shape" activity at the bottom of the *Shape Scavenger Hunt* activity sheet. Have students revisit their mystery shapes either by using the photograph they took of the shape or by returning to the place that they found the shape. Then allow students to use the Internet to research their shape (e.g., name and characteristics).

Part Two
Explaining the Concept

1 Tell students that they are going to further investigate the concepts of area and perimeter by finding the area and perimeter of irregular two-dimensional shapes. Ask students to share in their own words the definition of a regular shape and an irregular shape. Discuss the differences between irregular and regular shapes.

2 First model for students how to calculate the perimeter of irregular shapes using their TI-15. Draw an irregular shape on the board or find an example of a two-dimensional irregular shape either in the classroom, in magazines, in newspapers, or on the Internet. Preferably these shapes should be real-life objects. Help students measure the sides of the shape and find the sum of the sides.

3 On the board or overhead, draw the shape below and label its dimensions.

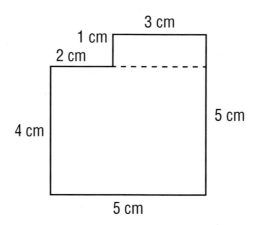

Shape Investigations (cont.)

Part Two
Explaining the Concept (cont.)

4 Show students how to use the measurements of the sides of the shape to calculate that the perimeter of the irregular shape is 20 centimeters. Have students press [4] [+] [2] [+] [1] [+] [3] [+] [5] [+] [5] [Enter].

5 Show students how to break an irregular shape into regular shapes. Then show students how to calculate that the area of the irregular shape is 23 centimeters by pressing: [(] [1] [×] [3] [)] [+] [(] [4] [×] [5] [)] [Enter]. Students can also determine the area by using the memory-storing function on their TI-15s by pressing: [1] [×] [3] [Enter] [▶M], [Enter] [4] [×] [5] [Enter] [+] [MR/MC] [Enter].

6 Have students work in pairs to calculate the perimeter and the area of the irregular shape shown below. *(perimeter = 28 cm; area = 41 cm)* When finished, review the answers as a class.

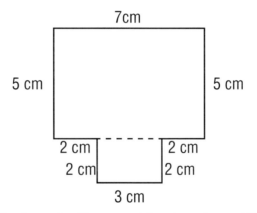

7 Distribute the *Measuring Irregular Shapes* activity sheet (page 160; page160.pdf) to students. Have them complete the activity sheet individually or in pairs. Then have students create a collage of five irregular shapes that they have found in books, magazines, newspapers, or on the Internet. They can either cut these shapes out or draw them by hand. Have students calculate the area and perimeter for each of these irregular shapes.

Applying the Concept

1 Tell students that they are going to be creating a design for a new park in Geometry City, called "Area and Perimeter Park." Brainstorm with the students the types of things that are found and needed in a city park.

2 Distribute the *Designing a Park* activity sheet (page 161; page161.pdf) to students. Tell them that they are to use regular and irregular polygons and/or two-dimensional shapes in their design to represent the items at the park. (Students do not need to draw the park with a scale or use realistic measurement for the design.)

Power Lesson 3

Shape Investigations (cont.)

Differentiation

- **Below Grade Level—** For the *Designing a Park* activity, have students work in pairs to create the design together. Provide the pairs with geometric cutouts that they can use on their design.

- **Above Grade Level—** For the *Designing a Park* activity, have students include a mystery three-dimensional shape in their park design. Have students find the surface area of this three-dimensional shape.

Applying the Concept (cont.)

3. Students must include the following list of items in the design for their parks:
   - at least three rides
   - one playing field
   - any necessary facilities, like restrooms, food stands, etc.
   - labels for the rides, playing fields, facilities, etc.
   - labels for the types of shapes used to represent the different parts of their design
   - labels of the measurements of the sides of the shapes used in their design
   - calculations of the area and perimeter of each shape used in the design

Extending the Concept

Have students create a three-dimensional design for Geometry City. Tell them that they are to use three-dimensional shapes in their design to represent locations in the city. Just like the *Area and Perimeter Park* activity, students do not need to draw the city with a scale or use realistic measurements for the design. Have students label the types of shapes used to represent the different parts of their city. Then have students calculate the surface area and volume of each shape used in the design of their cities.

Name _____

Power Lesson 3

Shape Scavenger Hunt

Directions: Complete the chart below and on the next page, then go on a scavenger hunt to find shapes around your school. In the last column, write the locations where you find each shape.

Three-Dimensional Shapes			
Shape Name	Picture	Characteristics	Location

© Shell Education #50616—30 Mathematics Lessons Using the TI-15 **157**

Shape Scavenger Hunt (cont.)

Two-Dimensional Shapes			
Shape Name	Picture	Characteristics	Location

Directions: Complete the chart using a shape that you do not know from around your school.

Mystery Shape		
Picture	Characteristics	Location

Power Lesson 3

Name _____

Investigating Shapes

Directions: Find a shape that you do not know the name of. Take a photograph of this mystery shape or visit the spot where you found the mystery shape. Follow the steps below and answer the questions to learn the name of your mystery shape.

1. On a separate sheet of paper, draw a congruent or similar picture of your mystery shape. Label it as *congruent* or *similar*.

2. Is your shape two-dimensional or three-dimensional? _____

3. If your shape is two-dimensional, how many sides and vertices does your shape have?

   _____

   If your shape is three-dimensional, how many faces and vertices does your shape have?

   _____

4. If your shape is two-dimensional, describe the relationships between the sides, e.g., *opposite sides are parallel*. If your shape is three-dimensional, describe the faces, e.g., *has a triangular base*.

   _____

5. Describe the other observations about your shape, such as congruence or types of angles.

   _____
   _____
   _____
   _____

6. What is the perimeter of your two-dimensional shape? _____

7. What is the area of your two-dimensional shape? _____

8. If you have a three-dimensional shape, what is the surface area? _____

9. If you have a three-dimensional shape, what is the volume? _____

10. Use the prefixes below to help predict the name of your shape. Remember, sides help identify two-dimensional shapes and bases help identify three-dimensional shapes.

    tri = three; quad = four; pent = five; hex = six; hept = seven; oct = eight

    I predict my shape is called _____.

© Shell Education   #50616—30 Mathematics Lessons Using the TI-15

Name _____

Measuring Irregular Shapes

Directions: The shapes below are all irregular. Use the measurements provided and your TI-15 to find the perimeter and area of each irregular shape.

1.

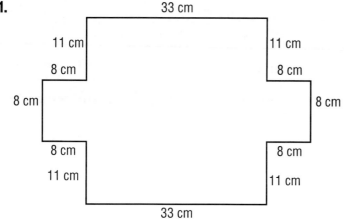

Perimeter: _____

Area: _____

2.

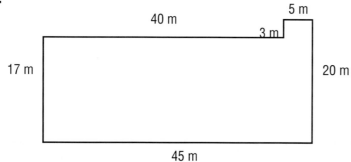

Perimeter: _____

Area: _____

3.

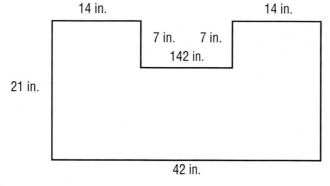

Perimeter: _____

Area: _____

160 #50616—30 Mathematics Lessons Using the TI-15   © Shell Education

Name _____

Designing a Park

Directions: Create a design for a new park in Geometry City called Area and Perimeter Park. Draw your design in the square below. Be sure to include everything listed in the checklist.

Checklist

- ☐ at least three rides
- ☐ one playing field
- ☐ facilities
- ☐ labels
- ☐ calculate area and perimeter of the shapes

Lesson 16

Hamster Haven

Overview

Students will construct cylinders without ends from rectangles, and then construct square prisms without ends to enclose them.

Mathematics Objective

Understands the basic measures of perimeter, area, and circumference.

TI-15 Functions

- Fractions
- Memory
- Operations

Materials

- TI-15s
- construction paper (large sheets)
- sheets of $8\frac{1}{2}$" x 11" paper
- centimeter ruler
- tape, rulers, scissors
- paper towel or toilet paper tubes
- *Harvey the Hamster* activity sheet (pages 165–166; page165.pdf)
- *Four Walls* activity sheet (page 167; page167.pdf)

Vocabulary

Complete the *Chart and Match* (page 19) vocabulary activity using the words below. Definitions for these words are included on the Teacher Resource CD (glossary.pdf).

- area
- circle
- centimeters
- circumference
- cylinder
- diameter
- dimensions
- height
- pi
- radius
- square centimeters
- width

Warm-up Activity

1. Give each student a piece of $8\frac{1}{2}$" x 11" paper. Ask students how to find the area of the piece of paper.

2. Ask them how to multiply $8\frac{1}{2}$" x 11". Have them use their TI-15s to multiply using the fraction keys:  .

3. The display shows $93\frac{1}{2}$. Have them press the [F↔D] to view the answer as a decimal.

4. Ask them what unit of measure to use with the $93\frac{1}{2}$. Have them draw a few horizontal and vertical lines along the edges of their papers to illustrate square inches.

162   #50616—30 Mathematics Lessons Using the TI-15   © Shell Education

Hamster Haven (cont.)

Explaining the Concept

1. Tell students you are going to make a cylinder for a hamster to play in. Ask students: *How could we make a cylinder?*

2. Give students paper cylinders such as toilet paper tubes or paper towel tubes. Ask students: *What shape is at the end of the cylinder?* Have students cut open the cylinders to create a net. Ask: *What shape is the net? How can we find the area of a paper towel tube using the net? How is the cylinder of a soda can different from a paper towel tube? How would their nets be different?*

3. Have students create their own cylinders. Give students an $8\frac{1}{2}$" x 11" (21.5 x 28 cm) piece of paper. Discuss the measurements of the paper. Have students position the paper in portrait style on their desks. Have them measure 1 cm from each vertical edge (long side of the paper) and place a hash mark for each measurement on the top and bottom of the paper. Using the hash marks and a ruler, have them draw a vertical line along each vertical edge to connect each pair of hash marks. Ask them to find the area of the rectangle between the two vertical lines. Ask: *If you roll the rectangle into a cylinder, what will the area be?* Have them roll the cylinders to overlap the vertical edges at the 1 cm lines.

4. Have students brainstorm how they could find the area if they could not cut open the cylinder and use the net. Ask: *What is the width of the rectangle represented on the cylinder?* (circumference) Ask: *What is the length of the rectangle represented on the cylinder?* (height of the cylinder)

5. Model how to calculate the area of cylinders. Ask students to identify and measure the diameter of the circle on the base of their cylinders. Give students the formula for finding the circumference: π x diameter or (π)d.

6. Investigate the meaning of *pi*. Have students divide the measurement of the circumference with the measurement of the diameter for the circular base of their cylinder. Tell students that if they could measure perfectly, they would always get the same number for the quotient of the circumference divided by the diameter. Tell them that the number is about 3.14 or π, pronounced "pie."

7. Show students the **π** key on the TI-15. Ask them to use it to find the circumference of a circle with a diameter of 7.2 cm by multiplying π times 7.2. It is about 22.62 cm. Ask them to repeat the exercise using 7 cm. The TI-15 will display 7π. Tell the students to press the [F↔D] key to see the decimal.

8. Distribute copies of the *Harvey the Hamster* activity sheet (pages 165–166; page165.pdf) and pieces of construction paper to students. Have them complete the activity.

© Shell Education   #50616—30 Mathematics Lessons Using the TI-15   163

Lesson 16

Hamster Haven (cont.)

Applying the Concept

1. Ask students how to make open boxes using four rectangles. Ask them how they could make open boxes that would just fit around the cylinder they made in the *Harvey the Hamster* activity.

2. Distribute copies of the *Four Walls* activity sheet (page 167; page167.pdf), and have students work on them.

3. When the students have finished making their open boxes and have placed their cylinders inside the boxes, have them draw what the open end looks like.

4. Ask them to pretend that the cylinder fits exactly into the box. Ask them to find the perimeter of the square.

Differentiation

- **Below Grade Level**—Have students use grid paper to count the number of square feet needed to cover the floor. Then, have them make their figures out of centimeter grid paper to make the lengths and areas obvious.

- **Above Grade Level**—Have students investigate filling cylinders that have bottoms with a liquid. Have them measure a graduated cylinder measuring in milliliters, comparing the area of the circle that is the base with the volume. Tell the students that a milliliter is equal to a cubic centimeter. Ask them to find a formula for the volume of a cylinder.

Extending the Concept

Have students design "doors" for their tubes and boxes and find the total area of their figures. Discuss the meaning of *surface area*.

164 #50616—30 Mathematics Lessons Using the TI-15 © Shell Education

Name _____

Harvey the Hamster

Directions: Read the text and answer the questions.

Jonah's little sister got a new pet hamster. To amuse the hamster, Jonah put the cardboard tube from a roll of paper towels in the cage. Harvey, the hamster, loved it. Jonah's sister did not like it. Jonah volunteered to cover the tube with flowered shelf paper. Jonah measured the diameter of the bottom of the tube. It was 4 cm. He measured the length of the roll. It was 28 cm.

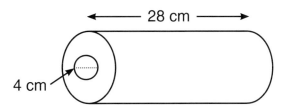

Jonah tried to imagine what the cardboard cylinder would look like if he could cut it apart. He thought it would look like this.

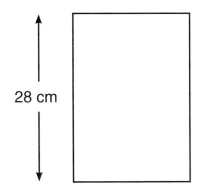

He knew that the height of the rectangle would be the same as the height of the roll of paper towels, but he did not know what to do about the width.

1. What is the shape at the bottom of the cylinder? _____

2. Where on the roll does the width of the rectangle come from? _____

3. What do you call the distance around a circle? _____

4. How do you find the circumference of a circle? _____

5. What is the width of the rectangle? _____

6. What is the area of the rectangle? _____

Lesson 16

Harvey the Hamster (cont.)

Directions: Read the text and answer the questions.

Harvey was happy until he grew. Soon, he did not fit into his tube. Jonah decided to make Harvey a new tube out of a piece of transparent plastic. His plan was to make a rectangle and glue the edges together. To find out the size of the cylinder, he managed to wrap a string around Harvey's chubby body and found that he was 20 cm around. Jonah decided to use trial and error to find a diameter for the cylinder that Harvey could fit into. He is going to keep the length at 28 cm.

7. Use these diameters to find the circle that Harvey will fit into. Find the area of the rectangle.

Diameter	Circumference	Area	Diameter	Circumference	Area
4.5 cm			6.5 cm		
5 cm			7 cm		
5.5 cm			7.5 cm		
6 cm			8 cm		

8. Which one would you choose for a growing hamster? _____

9. What is the circumference of the circle? What will the width of the rectangle be?

   _____

10. Jonah will need to add 0.5 cm on each end of the rectangle to have enough plastic to overlap so that he can glue it. What is the total width of the rectangle?

    _____

11. What is the area inside of the plastic cylinder? _____

12. Use construction paper and tape to make a model of Harvey's home.

Name _____

Four Walls

Lesson 16

Directions: Read the text and answer the questions.

Harvey did not like the transparent tube because of the light. However, it was good because it could be washed. Jonah decided to enclose his tube inside an open box that just fits around it, so that the plastic cylinder can be removed and washed.

1. Trace a circle around the end of your cylinder. What is the diameter of the circle? Draw the smallest square you can around the circle. What is the length of the side of the square?

You are going to make a box with a length of 28 cm, with openings at the ends that are the size of the square.

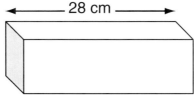

2. What is the shape of one of the sides? Sketch it on a separate sheet of paper.

   _____

3. What is the length of one of the rectangles? _____

4. What is the height of one of the rectangles? _____

5. What is the area of one of the rectangles? _____

6. How many rectangles are there? _____

7. What is the area of all the rectangles? _____

8. Each rectangle needs to be 0.5 cm wider on each end so that you have enough room to fold the ends over and tape them together. Sketch the rectangle on the sheet of paper.

9. How wide will each rectangle be? _____

10. What is the new area of each of the rectangles? _____

11. Cut out the rectangles and tape them together to make the box. Slide the cylinder inside the box. Which figure took more material to make?

    _____

© Shell Education #50616—30 Mathematics Lessons Using the TI-15 **167**

Lesson 17

Measuring Mania

Overview

Students will use measuring tools and calculators to find pairs of measurements. Then, they will use the measurements to make conclusions about different units of measure.

Mathematics Objective

Knows approximate size of basic standard units and relationships between them.

TI-15 Function

- Operations

Materials

- TI-15s
- rulers, yardsticks, or tape measures
- chalk or masking tape
- tape measures
- *Use Your Feet!* activity sheet (page 171; page171.pdf)
- *Measure That Rectangle* activity sheet (pages 172–173; page172.pdf)

 Vocabulary

Complete the *Total Physical Response (TPR)* (page 18) vocabulary activity using the words below. Definitions for these words are included on the Teacher Resource CD (glossary.pdf).

- area
- parallel lines
- perimeter
- rectangle
- square

Warm-up Activity

1. Ask students if they have ever watched the Olympic Games. Ask them what their favorite sports are.

2. Ask students what units of measure are used for the distances in the races.

3. Ask students if they have ever heard about horses being measured using "hands." A hand is four inches. The story behind this unit of measurement is that a child wanted to measure his favorite horse. Not having a device to measure with, he used the only thing he knew would be consistent: the palm of his hand. Since then, the hand has been the unit of measure for horses. The letters *HH* or *hh* after the numbers stand for "Hands High." The single letter *H* or *h* may be used, representing "hands."

4. Have students discuss why different units of measure are used for the same thing.

5. Have students measure their own height using their TI-15s. Have students break into pairs and count how many calculators tall they are. Have each pair share their height in TI-15s with the class.

Measuring Mania (cont.)

Explaining the Concept

1 Explain to students that they are going to investigate different units of measurement to see how they are related. Tell students they will be measuring rectangles. They will be finding the area and perimeter using standard and nonstandard units of measurement.

2 Review the concept of parallel lines. Have students demonstrate knowledge of this concept by using their arms to represent such lines. Remind students that the sides of a rectangle are parallel and of equal length.

3 Review the concept of calculating the area of a rectangle. Go over the formula with students ($A = l \times w$). Show students how they can use their TI-15s to find the area of a rectangle. Draw a rectangle on the board or overhead with a length of 10 inches and a width of 5 inches. Have students press [1] [0] [×] [5] [Enter]. The area of the rectangle is 50 square inches.

4 Have students review the concept of the perimeter of a rectangle. Review the formula $P =$ *the sum of all sides*. Ask students how they could use their TI-15s to calculate the perimeter of a rectangle. What operation key would they use? Have students use their TI-15s to find the perimeter of the rectangle used in Step 3. Have students press [1] [0] [+] [5] [+] [1] [0] [+] [5] [Enter]. The perimeter of the rectangle is 30 inches.

5 Divide the class into three groups. Draw four different-sized rectangles in chalk outside or on the floor inside using masking tape. Distribute copies of the *Use Your Feet!* activity sheet (page 171; page171.pdf) to students. Each group will need a tape measure. Tell students that they will measure four rectangles. They will use both standard units of measurement (inches) and nonstandard units of measurement (their feet). Each student in each group will need to use his or her own feet to measure the rectangle. Show students how to walk toe to heel along the rectangle and count their steps.

6 Allow students to work as a group to measure the rectangle using the tape measure and their feet. When all measurements have been recorded, the groups will rotate to a different rectangle. Have groups continue to rotate until each group has measured every rectangle.

7 When the class has completed their measurements, have students compare their results. Ask students of differing foot sizes to share their results. Discuss why the tape measurements are consistent while the feet measurements are not. Students should then complete the activity sheet.

Lesson 17

Measuring Mania (cont.)

Applying the Concept

1. Distribute copies of *Measure That Rectangle* (pages 172–173; page172.pdf). Have students measure the rectangles with the width of two of their fingers and centimeters on a ruler.

2. Ask students to take very precise measurements, using decimals where necessary.

Differentiation

- **Below Grade Level—** Have students measure in inches rather than centimeters to make the numbers they are working with smaller.

- **Above Grade Level—** Have students measure the width of their two fingers in centimeters. Then have students adjust their measurements in the chart accordingly. Were they able to match the measurements they took with the tape measure?

Extending the Concept

Ask students if they found measuring and/or computation easier with inches or centimeters. Ask them which one they think is more accurate. Student opinions may vary. Insist that students support their opinions.

Name _____

Use Your Feet!

Lesson 17

Directions: Follow the steps below.

Step 1 Measure the **length** of the rectangle using your feet. Record your measurement in the first table below.

Step 2 Measure the **width** of the rectangle using your feet. Record your measurement in the first table below.

Step 3 With your group, measure the **length** of the rectangle in inches using your tape measure. Record your measurement in the second table below.

Step 4 With your group, measure the **width** of the rectangle in inches using your tape measure. Record your measurement in the second table below.

Rectangle	Length (foot)	Width (foot)
1		
2		
3		
4		

Rectangle	Length (tape measure)	Width (tape measure)
1		
2		
3		
4		

Directions: Using the measurements above and your TI-15, find the area and perimeter for each rectangle. Write your answers in the charts below.

Rectangle	Perimeter (foot)	Area (foot)
1		
2		
3		
4		

Rectangle	Perimeter (tape measure)	Area (tape measure)
1		
2		
3		
4		

1. Compare the measurements you took with your feet with those that other members of your group took. Are all the measurements the same? Why or why not?

_____

2. Suppose a man with very large feet measured a rectangle with his foot. Would he get a larger or smaller number than you did? How do you know?

_____

Name _____

Measure That Rectangle

Directions: Measure each rectangle below using the width of two of your fingers. Then measure each rectangle again in centimeters using a ruler. Record your measurements on the next page.

1.

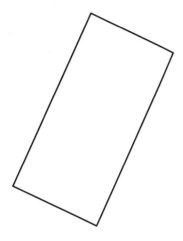

2.

3.

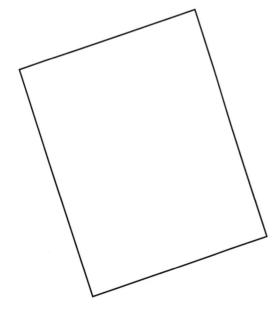

4.

Measure That Rectangle (cont.)

Directions: Record your measurements for each rectangle and answer the questions.

Rectangle	Length (fingers)	Width (fingers)
1		
2		
3		
4		

Rectangle	Length (ruler)	Width (ruler)
1		
2		
3		
4		

Directions: Using the measurements above and your TI-15, find the area and perimeter for each rectangle. Write your answers in the charts below.

Rectangle	Perimeter (fingers)	Area (fingers)
1		
2		
3		
4		

Rectangle	Perimeter (ruler)	Area (ruler)
1		
2		
3		
4		

5. Compare your measurements with a partner. Are your measurements the same? Why or why not?

_____

6. Why are methods of measurement standardized? Why can't we use our fingers to calculate the exact perimeter or area of a rectangle?

_____

7. Give an example of when we can use nonstandard units of measurement to estimate perimeter and area.

_____

_____

Lesson 18

Cover It Up

Overview

Students will compute areas of rooms and calculate the cost of purchasing and installing different kinds of floor coverings.

Mathematics Objective

Understands the basic measures of perimeter, area, volume, capacity, mass, angle, and circumference.

TI-15 Functions

- Fractions
- Memory
- Operations
- Parentheses

Materials

- TI-15s
- rulers
- *Beneath the Feet* activity sheet (pages 177–178; page177.pdf)
- *Kitchen Floor* activity sheet (page 179; page179.pdf)

 Vocabulary

Complete the *Music Makers* (page 18) vocabulary activity using the words below. Definitions for these words are included on the Teacher Resource CD (glossary.pdf).

- area
- dimensions
- square foot
- square inch

Warm-up Activity

1. Ask students how you the find area of a rectangle. Place students in groups of four.

2. Have groups find four different rectangles around the room, and measure and record their lengths and widths.

3. Instruct groups to use their TI-15s to calculate the areas of the rectangles they chose around the room.

4. As a class, discuss how area changes as the dimensions of the rectangles increase or decrease.

Cover It Up (cont.)

Explaining the Concept

1 Tell students that they are going to find the cost of replacing flooring in a home. Ask them what kinds of flooring are common in a home. Explain that most homes will have carpeting, hardwood, laminate (manufactured wood), ceramic tile, or vinyl flooring.

2 Explain to students that carpeting is normally sold by the square foot, hardwood and laminate may be sold by the square foot or by the board, ceramic tile may be sold by the tile or square foot, and vinyl flooring is sold by the square foot in strips. Explain that when you purchase flooring, the cost of installation is extra and must be figured into the total cost.

3 Tell students that the first room is 11 x 7 feet. It will be carpeted with carpeting that costs $10.50 per square foot, and with installation costing $1.50 per square foot. Ask students to find several ways to use the TI-15 to find the total cost of the job.

4 Students may find the total by finding the cost of the surface, writing it down, finding the cost of the installation, and then adding the cost of the surface. Work through this computation with the students.

5 Suggest that finding the cost of the surface and storing it in the memory may be helpful. To do this, enter the cost of the surface, press [Enter] [▶M] and [Enter], find the cost of installation, then add the number in memory by pressing [+] [MR/MC] and [Enter]. Work through this computation with students.

6 Students may wish to complete the computation in one step by adding the cost of the carpeting and installation in parentheses and multiplying by the number of square yards. Work through this computation with the students.

7 Ask students how they would find the area if the dimensions of the room were 15' 7" x 20' 6". Show them that they could write the inches as parts of a foot. 15' 7" would be $15\frac{7}{12}$ feet and 20' 6" would be $20\frac{6}{12}$ feet. (Fractions could be simplified if you wish, but it is not necessary.) Multiply the mixed numbers as a class, to find the area. To do this, have students press [1] [5] [Unit] [7] [n] [1] [2] [d] [×] [2] [0] [Unit] [6] [n] [1] [2] [d] [Enter].

8 Then, show students how to change the area to a decimal and store it in the memory by pressing [F↔D] [▶M] [Enter].

9 Distribute copies of the *Beneath the Feet* activity sheet (pages 177–178; page177.pdf) and have students complete them in small groups.

Lesson 18

Cover It Up (cont.)

Applying the Concept

1. Ask students to consider covering a floor with tiles. Ask them what they would need to know to decide how many tiles are required. Make sure they understand that they would need to know the shape and size of the tiles.

2. Distribute copies of the *Kitchen Floor* activity sheet (page 179; page179.pdf) and have students complete them. Tell students to round the answer up because they cannot buy part of a tile.

Differentiation

- **Below Grade Level—** Have students use grid paper to count the number of square feet needed to cover the floor. Show them that adding the cost of one square foot for each square on the grid is the same as multiplying by the number of squares.

- **Above Grade Level—** Have students calculate the cost of flooring for irregular-shaped rooms by giving them the dimensions of rooms with unusual geometric shapes.

Extending the Concept

Have students consider laying vinyl flooring that can only be purchased in strips that are 25 feet long and 3 feet 2 inches wide. Look at rooms whose dimensions are more or less than 25 feet long and decide the most economical way to lay the flooring.

Name _____

Beneath the Feet

Directions: Read the text carefully. Answer the questions below. Round all questions to the nearest cent.

Kimi's mother is planning to replace the floor covering in several rooms in the house. She asked Kimi to estimate the cost. They are considering several different kinds of floor coverings. The possibilities are listed below.

Surface	Cost	Installation
carpeting	$12.74 per sq. ft.	$1.38 per sq. ft.
hardwood	$4.20 per sq. ft.	$4.49 per sq. ft.
laminate	$2.52 per sq. ft.	$3.76 per sq. ft.

Kimi decided to start with the family room. It is rectangular and measures 10 x 14 feet.

1. What is the area of the family room? _____

Kimi decided to find the total by finding the cost of the surface, writing it down, finding the total cost of the installation, and then adding the two together.

2. Using this method, what is the total cost of installed carpeting? _____

Kimi thought it might be more efficient to find the cost of the surface, store it in the memory by pressing [▶M] [Enter], find the cost of installation, and add it to the number in the memory by pressing [+] [MR/MC] [Enter].

3. Using this method, what is the total cost of installed hardwood? _____

Kimi remembered that it is possible to do the computation in one step by using parentheses.

4. What two numbers will be added in the parentheses for the laminate?

_____

5. Write the computation she will perform to find the total cost of laminate.

_____

6. Using this method, what is the total cost of installed laminate? _____

7. Which method of computation did you find to be the easiest? Explain your reasons.

_____

_____

Beneath the Feet (cont.)

Directions: Read the text carefully. Answer the questions below.

Kimi decided to find the estimates for her own bedroom next. She has the same choices listed in the chart on the first page. The dimensions of her room are 10' 3" x 12' 6".

8. Why is it more difficult to find the area of Kimi's room than it was to find the area of the family room?

   _____

9. How many inches are in a foot? _____

10. How could you write 10' 3" only in feet using a fraction? _____

11. How could you write 12' 6" only in feet using a fraction? _____

12. To find the area of Kimi's room, use the fraction keys to perform the multiplication. On your TI-15, press: [1] [0] [Unit] [3] [n] [1] [2] [d] [×] [1] [2] [Unit] [6] [n] [1] [2] [d] [Enter]. What is the answer?

    _____

13. Use the [F↔D] key to change the answer to a decimal. Press [▶M] [Enter] to store the number in memory. Use parentheses and the number in memory to find the total cost of installing and carpeting Kimi's room. Press [(] [1] [2] [.] [7] [4] [+] [1] [.] [3] [8] [)] [×] [MR/MC] [Enter].

    What is the total cost of carpeting? _____

14. Find the cost for hardwood. _____

15. Find the cost for laminate. _____

16. What would you choose for your own room? Why?

    _____

    _____

    _____

Name _____

Kitchen Floor

Directions: Read the text carefully. Answer the questions below.

Kimi's mother wants to use ceramic tile in the kitchen. The dimensions of the kitchen are 16 x 12 feet. For tile #1, each tile is a square with sides that are 6" long. For tile #2, each tile is a square with sides that are 12" long. The costs are shown below.

Surface	Cost	Installation
Ceramic Tile #1	$2.31 per 6" square tile	$9.07 per sq. ft.
Ceramic Tile #2	$4.36 per 12" square tile	$9.07 per sq. ft.

1. Which tile do you think will be cheaper? _____

2. What is the area of the kitchen in square feet? _____

3. Multiply each dimension to the kitchen by 12 to convert its dimensions to inches. What is the area of the kitchen in square inches? _____

Use the information about Tile #1 to answer the questions below.

4. What is the area of a tile in square inches? _____

5. How many tiles are needed for the kitchen? _____

6. What is the cost of the tiles? _____

7. What is the cost of installation? _____

8. What is the cost of using Tile #1? _____

Use the information about Tile #2 to answer the questions below.

9. What is the area of a tile in square inches? _____

10. How many tiles are needed for the kitchen? _____

11. What is the cost of the tiles? _____

12. What is the cost of installation? _____

13. What is the cost of using Tile #2? _____

14. Which tile is cheaper? _____

Lesson 19

Investigating Volume

Overview

Students will learn how to calculate volume using linking cubes.

Mathematics Objective

Understands the basic measure concept of volume.

TI-15 Function

- Operations

Materials

- TI-15s
- 1-inch linking cubes
- boxes of varying sizes (e.g., cereal, tissue, cracker boxes)
- rulers
- materials for building boxes (cardboard, heavy paper, tape, scissors, etc.)
- *Boxes of Volume* activity sheet (page 183; page183.pdf)
- *Build It!* activity sheet (pages 184–185; page184.pdf)

 Vocabulary

Complete the *Chart and Match* (page 19) vocabulary activity using the words below. Definitions for these words are included on the Teacher Resource CD (glossary.pdf).

- area
- volume

Warm-up Activity

1. Review the formula for finding the area of a rectangle. (A = l x w)

2. Draw a rectangle on the board. Tell students that the length of the rectangle is 12 inches and the width is 8 inches. Ask them to calculate the area of the shape. Students can use the TI-15 to complete these calculations.

3. Hold up one of the boxes being used in today's activity. Ask students to estimate the length and width of the box. Have a student use a ruler to find actual measurements of the box. Compare the results with the class.

Investigating Volume (cont.)

Explaining the Concept

1 Ask students how they would define *volume*. (*The amount of space that the shape occupies.*) Tell students that they will be measuring volume by using units that also take up space, such as cubic inches. The number of cubic units of space that the shape occupies is the volume measurement of that shape.

2 Create a 3 x 4 x 2 inch rectangular prism and then display it for students or lay it on the overhead projector. Tell students that volume is the amount of space that "fills" a shape. Ask students: *If these are 1-inch cubes, what do you think the volume of this rectangular prism is?* This will help students to visualize the concept.

3 Ask students to imagine building layers of cubes on the base up to the height of the prism. This will help them see how volume is calculated.

4 Write the formula for volume (*V = l* x *w* x *h*) on the board or overhead. Work through the following example with students: *If the length of an object is 4 inches, the width measures 3 inches, and the height measures 2 inches, what is the volume?* Have students substitute these measurements into the formula. Have them use the TI-15 to find the volume by pressing [3] [x] [4] [x] [2] [Enter]. (*V = 24 cubic inches*)

5 Distribute copies of *Boxes of Volume* (page 183; page183.pdf) to students. Distribute 1 box to each student or pair of students. Be sure that these boxes are not too large as students must be able to fill the box with cubes. Make sure students have a sufficient number of cubes to complete the activity.

6 Ask students if all of the boxes are the same size. See if they can estimate how many 1-inch cubes can fit inside each of their boxes. Have them record their estimates on the activity sheet.

7 Have students place cubes inside their boxes. They will need to figure out how many cubes it will take to make the length of the box, the width of the box, and the height of the box. Have students record their findings on their activity sheets.

8 After students complete their activity sheets, be sure to discuss responses to the questions as a class. Also, discuss how the students' estimates compare with their actual findings.

Lesson 19

Investigating Volume (cont.)

Applying the Concept

1. Distribute copies of the *Build It!* activity sheet (pages 184–185; page184.pdf), and have students complete the activity.

2. Remind students that the volume of a rectangular prism can be found by multiplying the length and width of the prism to find the area of the base, then multiplying the area of the base by the prism's height.

Differentiation

- **Below Grade Level—** Provide students with two or three nets of rectangles so that they can visualize how to create a box. Students could also deconstruct cereal boxes to learn how to fold a box. Have students work in groups of three or four to complete the *Build It!* activity sheet.

- **Above Grade Level—** Ask students to investigate the concept of measuring liquid volume. They will need to investigate the number of cubic inches in a gallon. They will find it easier to work with metric units because 1 cubic centimeter is equal to 1 milliliter.

Extending the Concept

Have students use a ruler to measure and calculate the volume of larger boxes.

Have students explore the similarities and/or differences of finding the volume of a prism versus finding the volume of a cylinder.

Have students construct a model of square centimeters, square inches, square feet, or square meters.

Name _____

Boxes of Volume

Lesson 19

Directions: In the charts below, record your estimates and findings for each of your boxes.

Estimates	Box 1	Box 2
number of cubes for length of box		
number of cubes for width of box		
number of cubes for height of box		
volume (cubic inches)		

Actual	Box 1	Box 2
number of cubes for length of box		
number of cubes for width of box		
number of cubes for height of box		
volume (cubic inches)		

Directions: Answer the questions about your investigation.

1. How many 1-inch cubes did it take to fill up each box?

   _____

2. Does the number of 1-inch cubes used to fill each box match the volumes found for each box? Why or why not?

   _____

   _____

3. What does this tell you about the meaning of volume?

   _____

   _____

Name _____

Build It!

Directions: Answer the questions and follow the steps carefully.

You are going to build a box to store some of your school supplies. Use the available materials to build your own box. Before you begin, make a plan for your box. You will need to be realistic. Your box will not be able to hold everything. You will need to think about the amount of materials you have to build your box.

1. What objects will you put inside the box?

   _____

   _____

2. What are the dimensions of the objects that you will put in the box?

   _____

   _____

   _____

3. What is the approximate volume of the objects that you will put in your box?

   _____

   _____

4. Based on the dimensions and volume of the objects to be stored in the box, determine the dimensions of your box. Sketch a picture of your box below. Label the dimensions.

Build It! (cont.)

Directions: Answer the questions and follow the steps carefully.

5. What is the volume of the box? Will it hold the total volume of items in the box?

___

Directions: Now build your box using the available materials in the classroom and then answer the questions below.

6. Do all the objects that you listed in question 1 fit inside the box? Why or why not?

___

7. What challenges did you face in building your box? What things went smoothly?

___

8. Does your box look like your drawing? How is it similar and/or different?

___

9. What would you change about your box? What would you keep the same?

___

Lesson 20

Fish for the Science Fair

Overview

Students will learn to use their knowledge of fractions to convert units of measure using unit cancellation.

Mathematics Objective

Understands the relationships among linear dimensions, area, and volume, and the corresponding uses of units, square units, and cubic units of measure.

TI-15 Functions

- Conversions
- Fractions
- Operations

Materials

- TI-15s
- *Fish Frenzy* activity sheet (page 189; page189.pdf)
- *Fish Food* activity sheet (page 190–191; page190.pdf)

Vocabulary

Complete the *Chart and Match* (page 19) vocabulary activity using the words below. Definitions for these words are included on the Teacher Resource CD (glossary.pdf).

- capacity
- cup
- gallon
- ounce
- tablespoon
- teaspoon
- unit cancellation

Warm-up Activity

1. Tell students that 1 mile = 1.61 kilometers.

2. Tell them that the distance between Washington, D.C. and Los Angeles is about 2,700 miles.

3. Ask them to use their TI-15s to find out how many kilometers there are between Washington, D.C. and Los Angeles. *(4,347 km)*

4. Tell students that the distance between Paris, France and Berlin, Germany is about 877 km.

5. Ask them to find how many miles there are between Paris and Berlin. *(about 545 miles)*

Fish for the Science Fair (cont.)

Explaining the Concept

1 Tell students that it is important to understand how to cancel when multiplying fractions and converting measurements. Display the problem below. Tell students that the denominator of the second fraction can also be shown as *2 times 5* because 2 x 5 and 10 are equivalent. Remind students that multiplication and division are inverse operations. That means that multiplying and dividing by the same number can cancel out the effects of each other. Ask them which factors cancel each other. Show them that the fives cancel because one is a numerator and one is a denominator. The answer is $\frac{7}{2}$.

$$\frac{5}{1} \times \frac{7}{10} \qquad \frac{5}{1} \times \frac{7}{2 \cdot 5} \qquad \frac{\cancel{5}}{1} \times \frac{7}{2 \cdot \cancel{5}}$$

2 Show students the first problem from the warm-up exercise. Change the distance from Washington to Los Angeles from miles into kilometers. Set up the problem as seen below. Ask them to explain why $\frac{1.61 \text{ km}}{1 \text{ mile}}$ is another name for 1. *(They represent the same distance.)*

$$\frac{2{,}700 \text{ miles}}{1} \times \frac{1.61 \text{ km}}{1 \text{ mile}}$$

3 Show them that they can cancel the miles, just like they cancelled the fives in the first problem because miles is the same unit. Have them use their TI-15s to complete the calculation by multiplying 2,700 by 1.61.

$$\frac{2{,}700 \;\cancel{\text{miles}}}{1} \times \frac{1.61 \text{ km}}{1 \;\cancel{\text{mile}}}$$

4 Show the Paris to Berlin computation below. Ask students why 1 mile/1.61 km is another name for 1. Explain that 1.61 km is the denominator this time so that it will cancel out the km in the other numerator because it is the same unit. Ask students how to complete the computation. Have them divide 877 by 1.61 on their TI-15s.

$$\frac{877 \text{ km}}{1} \times \frac{1 \text{ mile}}{1.61 \text{ km}} \qquad \frac{877 \;\cancel{\text{km}}}{1} \times \frac{1 \text{ mile}}{1.61 \;\cancel{\text{km}}}$$

5 Tell students that this method of converting from one unit of measure to another is called *unit cancellation*. Distribute copies of the *Fish Frenzy* activity sheet (page 189; page189.pdf). In groups of three, have students complete the activity sheet. Circulate among the groups to help students as they work.

Lesson 20

Fish for the Science Fair (cont.)

Applying the Concept

1 Explain to students that unit cancellation can be used to convert measurements within the same system of measure. As an example, ask students how to use unit cancellation to convert $7\frac{1}{2}$ yards into inches.

2 Discuss with students that there is more than one method of conversion. One way is to multiply $7\frac{1}{2}$ yards by $\frac{36 \text{ inches}}{1 \text{ yard}}$. Another method would be to multiply $7\frac{1}{2}$ yards by $\frac{3 \text{ feet}}{1 \text{ yard}}$, then multiply $\frac{12 \text{ inches}}{1 \text{ foot}}$. Help students solve the computation on their TI-15s using their chosen computation. The answer is 270 inches.

3 Distribute copies of the *Fish Food* activity sheet (page 190–191; page190.pdf). Have students complete the activity sheet using unit cancellation. Assist them in computations by changing two units of measure.

Differentiation

- **Below Grade Level—** Emphasize multiplying by 1 by showing 1 pound and 16 ounces on a balance, or by pouring 8 ounces of water in a cup.

- **Above Grade Level—** Have students measure different lengths of lines drawn on the board or pieces of string of alternate lengths, using units of measure from both metric and standard systems. Have students compare the metric and standard measurements they found and determine the difference between them by using their TI-15s.

Extending the Concept

Have students investigate problems involving speed, time, and distance using unit cancellation. For example, to compute the distance traveled in 5 hours when going 60 miles per hour, write the product:

$$\frac{5 \text{ hours}}{1} \times \frac{60 \text{ miles}}{1 \text{ hour}}$$

$$\frac{5 \cancel{\text{ hours}}}{1} \times \frac{60 \text{ miles}}{1 \cancel{\text{ hour}}} = 300 \text{ miles}$$

Name _____

Fish Frenzy

Directions: Read the text carefully. Solve the problem using unit cancellation.

Teresa is setting up two fish tanks at the fair. Each tank can hold up to 30 gallons of water. The tanks will all hold different amounts of fish. Each fish weighs 16 ounces. She needs to figure out how much water to put in each tank without overflowing the water.

8 fish in Tank 1   16 fish in Tank 2

   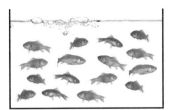

How much water should Teresa put in each tank? (Hint: 128 ounces = 1 gallon)

1. Start by figuring out how much all the fish weigh in each tank.

    a. 8 fish in Tank 1: $\frac{8 \text{ fish}}{1} \times \frac{16 \text{ oz.}}{1 \text{ fish}}$ = _____

    b. 16 fish in Tank 2: $\frac{16 \text{ fish}}{1} \times \frac{16 \text{ oz.}}{1 \text{ fish}}$ = _____

2. What unit of measure was cancelled out? _____

    What unit of measure are we left with? _____

3. Convert the total ounces into gallons.

    a. Tank 1: $\frac{\Box \text{ oz.}}{1} \times \frac{1 \text{ gallon}}{128 \text{ oz.}}$ = _____

    b. Tank 2: $\frac{\Box \text{ oz.}}{1} \times \frac{1 \text{ gallon}}{128 \text{ oz.}}$ = _____

4. What unit of measure was cancelled out? _____

    What unit of measure are we left with? _____

5. Now use your TI-15 to subtract the total gallons from 30 gallons (the capacity of the tank) to find out how much water each tank needs.

    Tank 1: _____

    Tank 2: _____

© Shell Education   #50616—30 Mathematics Lessons Using the TI-15

Lesson 20

Name _____

Fish Food

Directions: Read the text below. Convert the measurements, and show your work.

Teresa's science class is in charge of feeding the fish at the fair. Teresa's teacher, Mr. Pescadero, wants the class to make the fish food themselves in the science lab. Mr. Pescadero's recipe is below:

Veggie-Flakes Fish Food
(one jar makes about 17 oz.)
4 oz. wheat germ
6 oz. corn flour
6 oz. ground barley
2 tablespoons spinach powder
1 teaspoon water

However, Mr. Pescadero only has one teaspoon, and one measuring cup that measures $\frac{1}{8}$ cup, $\frac{1}{4}$ cup, $\frac{1}{2}$ cup, $\frac{3}{4}$ cup, and 1 cup. Help Teresa's class figure out how to convert the units of measure so they can complete the recipe. Use these equivalences to help you complete the conversions: 8 ounces = 1 cup; 1 tablespoon = 3 teaspoons; 48 teaspoons = 1 cup. Show your work.

1. Convert the wheat germ, corn flour, and ground barley from ounces to cups.

2. Convert the spinach powder from tablespoons to teaspoons.

3. How many cups of spinach powder are needed for the recipe?

Mr. Pescadero ordered all of the ingredients from the store. He came back to the science lab with the following amounts:

2-lb. bag of wheat germ
3-lb. bag of corn flour
3-lb. bag of ground barley
$\frac{1}{2}$-lb. bag of spinach powder

Fish Food *(cont.)*

Directions: Read the text below. Answer each question to complete the recipe.

4. One way to convert pounds into cups is to multiply the pounds by $\frac{16 \text{ oz.}}{1 \text{ lb.}}$, then multiply by $\frac{1 \text{ cup}}{8 \text{ oz.}}$. Find out how many cups of each ingredient Mr. Pescadero ordered.

   a. wheat germ: $\frac{2 \text{ lbs.}}{1} \times \frac{16 \text{ oz.}}{1 \text{ lb.}} = $ _____

   b. _____ $\times \frac{1 \text{ cup}}{8 \text{ oz.}} = $ _____

   c. corn flour: $\frac{3 \text{ lbs.}}{1} \times \frac{16 \text{ oz.}}{1 \text{ lb.}} = $ _____

   d. _____ $\times \frac{1 \text{ cup}}{8 \text{ oz.}} = $ _____

   e. ground barley: $\frac{3 \text{ lbs.}}{1} \times \frac{16 \text{ oz.}}{1 \text{ lb.}} = $ _____

   f. _____ $\times \frac{1 \text{ cup}}{8 \text{ oz.}} = $ _____

   g. spinach powder: $\frac{0.5 \text{ lb.}}{1} \times \frac{16 \text{ oz.}}{1 \text{ lb.}} = $ _____

   h. _____ $\times \frac{1 \text{ cup}}{8 \text{ oz.}} = $ _____

5. How many jars can Mr. Pescadero's class make?

   a. _____ cups of wheat germ total ÷ _____ cups needed = _____

   b. _____ cups of corn flour ÷ _____ cups needed = _____

   c. _____ cups of ground barley ÷ _____ cups needed = _____

   d. _____ cups of spinach powder ÷ _____ cups needed = _____

   e. Total jars: _____

Power Lesson 4

Easy as Pi

Overview

Students will investigate the value of π and the formulas for circumference and area.

Mathematics Objective

Understands the basic measures of perimeter, area, volume, capacity, mass, angle, and circumference.

TI-15 Functions

- Constant feature
- Fraction to decimal
- Operations
- π (Pi)

Materials

- TI-15s
- centimeter ruler, tape, string, scissors, sticky notes, sticker labels, and toothpicks
- circular lids and round objects
- *Parts of a Circle* activity sheet (page 197; page197.pdf)
- *Round About* activity sheet (page 198; page198.pdf)
- *Close Enough?* activity sheet (pages 199–201; page199.pdf)

Vocabulary

Complete the *Vocabulary Bingo* (page 19) vocabulary activity using the words below. Definitions for these words are included on the Teacher Resource CD (glossary.pdf).

- chord
- circumference
- diameter
- equilateral
- hexagon
- inscribed
- octagon
- pentagon
- perimeter
- radius
- ratio
- regular polygon
- square
- triangle
- vertex

Warm-up Activity

1. Ask students to describe a stop sign. Make sure they know that it is an 8-sided shape called an *octagon*.

2. Draw an octagon on the board with the sides of lengths of 7.45 inches, 9.94 inches, and 12.43 inches.

3. Ask students to find the perimeter of the octagon.

Easy as Pi (cont.)

Part One
Explaining the Concept

1 Tell students that they are going to compare and contrast the different parts of a circle and then build a three-dimensional model of those parts. Discuss with students the diameter, radius, chord, and circumference of a circle. Draw a large circle on the board or overhead with the center marked.

2 Then discuss the definition of a chord. *(a straight line on a circle joining two points on the circumference of the circle; it must pass through the center of the circle)* Have students identify different chords on the circle. Ask: *How do you measure the circumference of the circle?* Show them that it is very difficult to do. Tell them by the end of this lesson they will have learned an easier way to find the circumference of a circle.

3 Ask students: *How do you find the diameter of a circle?* They should know that the diameter is the longest chord that can be drawn across the circle. Hold one end of a piece of string and stretch it across the circle until you have reached the farthest distance. Ask students how to find the radius. Fold the string in half to show students the radius. Show students on the TI-15 how to divide the diameter by 2 to calculate the radius. Ask students to come up with two different ways to calculate the diameter using the operations and functions of the TI-15 (e.g., multiply by 2, add the radius twice, or store the operation x 2 into Op1).

4 Complete as a class the activity sheet *Parts of a Circle* (page 197; page197.pdf). Students will identify how the concept pairs in the activity sheet are alike and different. Do not complete the "Relationship" row of the activity sheet at this time. Make a transparency of the chart or draw the chart on the board. Give each pair of students a circular lid from a container, a ruler, and a piece of string to use as a model during the discussion. Students should use the model to explore the similarities and differences between the concepts as they are discussed.

5 Now have students use their circular lids to create a model of the parts of the circle. Provide students with a variety of supplies to create their model, such as labels that stick, string, sticky notes, toothpicks, and scissors. Students should label each of the parts with its name and measurement. Have students use their TI-15s to calculate the diameter and circumference.

6 Have students share their models in small groups or with a partner. When students share their work, others in the group should say one thing they like about the model and ask one question about the mathematics of the model.

7 Then demonstrate for students how to find the ratio of the circumference to the diameter. Tell students that they will be dividing the circumference measurements by the diameter measurements on their TI-15s by using the ÷ key. Tell students they will be rounding their answers to three decimals or to the thousandths place. Remind students they can press Fix and 0.001 if they would like the TI-15 to do the rounding for them.

© Shell Education   #50616—30 Mathematics Lessons Using the TI-15   **193**

Easy as Pi (cont.)

Part One
Explaining the Concept (cont.)

8 Discuss measuring errors with students. Make sure that students know how to use a centimeter ruler. Explain that they can estimate measurements as close as the hundredths place by estimating how far between the millimeter marks a measurement lays.

9 Tell students that they are going to average their group's findings to try to even out the errors. Explain that to find the averages, they will need to add their measurements together for each object and divide it by the number of members in their groups. Use the following example to remind students how to do this. Have students press [3] [.] [6] [3] [2] [+] [5] [.] [2] [4] [5] [+] [6] [.] [3] [2] [5] [+] [8] [.] [3] [6] [9] [Enter]. Then have students press [÷] [4] [Enter]. (5.893).

Applying the Concept

1 Distribute copies of the *Round About* activity sheets (page 198; page198.pdf). Place students into groups and have each group member measure the same four objects and record the information on his or her activity sheet. Tell them that they will need the circles and their *Round About* activity sheets later in this lesson.

2 After the class has finished the sheet, ask each group the number that it found for the average of the ratios. Discuss that you would expect them all to be about the same if all of the shapes were circular. If they were not the same, some of the objects could have been oval instead of round.

3 As a class, complete the "Relationship" section of the activity sheet *Parts of a Circle* (page 197; page197.pdf). You may also want students to go back to their models and use sticky notes to identify relationships between the parts of the circle.

Part Two
Explaining the Concept

1 Ask students to name some polygons and the number of sides that each has. Be sure that triangles (3 sides), rectangles (4 sides), pentagons (5 sides), hexagons (6 sides), and octagons (8 sides) are included. Ask students if there is a name for every possible kind of polygon. Explain that many-sided polygons can have a name like a 17-gon, which has 17 sides.

2 Tell students that a *regular* polygon is one that has all sides of the same length. Ask them another name for a regular polygon. (*square*) Explain that a regular triangle is called an equilateral triangle. Ask students: *How do you find the perimeter of a regular polygon if you know the length of one side?* Make sure that students understand that to find the perimeter you can multiply the number of sides by the length of the one side, since all sides are equal in length.

Easy as Pi (cont.)

Part Two
Explaining the Concept (cont.)

3. Distribute copies of the *Close Enough?* activity sheet (pages 199–201; page199.pdf). Explain that the two circles are the same size and the polygons are inscribed in the circles. Explain that *inscribed* means that each of the vertices touch the circle and that the polygons have the same center as the circle.

4. Explain what the radius of a circle is. Tell students the radius is half of the diameter. Review the formula $d = 2r$. Have students measure the radius of one of the circles on the activity sheet and multiply it by two on their TI-15s to find the diameter. Remind students the two circles are the same size, so they only need to measure one.

5. Have students locate each of the polygons on the circles. The first circle has the triangle, pentagon, and octagon. The second has the square, hexagon, and 17-gon. Be sure that they can locate each polygon and a side of the polygon to measure. Tell them to measure as accurately as possible. Try to measure from the middle of one point to the middle of the next one. Discuss why accurate measurement becomes more important for the shape with the larger number of sides. Work as a class to answer questions 1–4 on the activity sheet.

6. Tell students the number they have been investigating in the *Round About* activity sheet and this activity sheet, the ratio of the circumference of a circle to its diameter, is π, a Greek letter pronounced like "pie." It is the same for every circle and can also be found in other kinds of mathematics besides geometry. It is a number that never comes out even and its decimal places never form a repeating pattern.

7. Show students the [π] key on their TI-15s. Explain that this symbol stands for the exact ratio between the circumference and diameter of a circle. Show them that if they press [π] and [Enter] the display shows the symbol. If they then press [F↔D], a decimal approximation of π is displayed.

8. Show students how to use the [π] key in a computation by pressing it instead of a number. Explain that they have now found that $π = C ÷ d$. Ask them to find the related multiplication sentence. If necessary, show them that the related multiplication sentence for $4 = 12 ÷ 3$ is $12 = 4 × 3$. The formula for circumference is $C = π · d$.

Easy as Pi (cont.)

Differentiation

- **Below Grade Level—** Enlarge the two circles on the *Close Enough?* activity sheet so that students have a larger visual to examine. Also have students use different colored pencils to trace the perimeters of each of the inscribed circles.

- **Above Grade Level—** Have students test a circle with a larger diameter to see if their findings from the *Close Enough?* activity sheet hold true.

Applying the Concept

1. Have students complete problem 5 on the activity sheet *Close Enough?* independently. Remind them that they will need the circles they used in Part One.

2. Circulate around the room to assess student understanding and answer any questions students may have. When students have finished their activity sheets, go over the answers as a class.

Extending the Concept

- Explain that the number π occurs in areas of advanced mathematics in addition to geometry and that mathematicians have calculated π to many decimal places. Have the students research the value of π.

- Celebrate π Day. Ask the students why the time 1:59 and 26 seconds on March 14 is special to mathematicians. (*matches sequence of π: 3.1415926*)

- Have students find the formulas for the surface area $[A = 4\pi r^2]$ and volume $[V = (\frac{4}{3})\pi r^3]$ of a sphere and find the surface area and volume of a basketball.

Name _____

Parts of a Circle

Directions: Look at each concept pair below. Write about how each pair of words is alike and different. Then, describe the mathematical relationship between the words in each pair.

Concept Pair	Concept Pair	Concept Pair
radius diameter	diameter circumference	chord diameter
Alike:	**Alike:**	**Alike:**
Different:	**Different:**	**Different:**
Relationship:	**Relationship:**	**Relationship:**

Power Lesson 4

Name _____

Round About

Directions: With your group, measure four round objects with different diameters. Then, follow the instructions below.

1. Set the object on a piece of paper and trace around the object. Wrap a piece of string around the object, and then measure the length of the string. What measure of the circle will you be finding?

   _____

2. Stretch the string across each circle to find the longest distance possible. Measure the distance on the string. What measure of the circle will you be finding?

   _____

3. Measure each of your four objects in the same way and write the measurements below.

Object	Circumference	Diameter	Ratio

4. Find the ratio of the circumference to the diameter of each object. Round the answers to three decimal places and write them in the chart above. Save the objects.

You and your team have measured the same four objects. To even out measuring differences, find the averages of your measurements for each object and write them on the chart. Calculate the ratio of the circumference to the diameter for the averages.

Object	Circumference	Diameter	Ratio

5. Compare the average of your ratios with the rest of the class. Are they similar? _____

6. Calculate the class average of the ratios. Do you think that everyone would get the same ratio if it were possible to measure carefully?

   _____

Name _____

Close Enough?

Directions: Read the text, follow the instructions, and answer the questions.

The figure on the left includes a circle, an equilateral triangle, a regular pentagon, and a regular octagon. The figure on the right includes a circle, a square, a regular hexagon and a regular 17-gon. The circles are the same size and all the other figures are centered at the center of the circle.

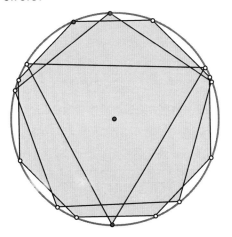
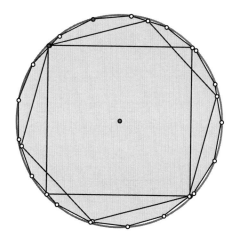

1. What is the radius of either of the circles in centimeters? _____

2. What is the diameter of the circles? _____

3. Find the length of the side and the perimeter of each of the figures. Calculate the ratio of the perimeter to the diameter of the circle. Enter the information into the chart below. Round to the nearest tenth of a centimeter.

Figure	Length of Side	Perimeter	Ratio
triangle			
square			
pentagon			
hexagon			
octagon			
17-gon			

Close Enough? (cont.)

Directions: Follow the instructions below.

4. Use the chart on the bottom of the previous page to answer the following questions.

   a. Which polygon had the smallest perimeter? _____

   b. Which polygon had the largest perimeter? _____

   c. Which perimeter would be the closest to the circumference of the circle?
   _____

   d. What would be true about the number of sides of a polygon whose perimeter was even closer to the circumference of the circle?
   _____
   _____
   _____

   e. Suppose you could keep drawing regular polygons with more and more sides. What would it look like? Why?
   _____
   _____
   _____

   f. What measure of the circle would the perimeter of the polygons get closer and closer to as you increase the number of sides?
   _____
   _____

   g. What happens to the ratio of the perimeter to the diameter as the number of sides increases?
   _____
   _____

Close Enough? (cont.)

5. Press π and Enter. What do you see?

_____

6. To see a decimal approximation for π, press π Enter and F↔D. What is the approximation for π used by the TI-15?

_____

7. Use the formula, $C = \pi \cdot d$ to find the circumferences of the circles on the first page of this activity sheet. Remember to use the F↔D and π keys.

8. Return to the *Round About* activity sheet and use the π key to calculate the circumferences of the objects you measured. Complete the chart below.

Object Measured	Measured Circumference	Calculated Circumference

Lesson 21

In the Long Run

Overview

Students will learn to find experimental and theoretical probability by flipping coins and tossing number cubes.

Mathematics Objective

Uses basic sample spaces to describe and predict events.

TI-15 Function
- Fractions

Materials
- TI-15s
- coin for each student
- number cube for each student
- paper cup for each student
- *Heads or Tails* activity sheet (pages 205–206; page205.pdf)
- *Fair Play* activity sheet (page 207; page207.pdf)

 Vocabulary

Complete the *Vocabulary Bingo* (page 19) vocabulary activity using the words below. Definitions for these words are included on the Teacher Resource CD (glossary.pdf).

- experimental probability
- frequency
- outcome
- prediction
- probability
- random
- tally sheet
- theoretical probability
- trial

 Warm-up Activity

1. Ask each student to raise a hand. Have students who raised their right hands stand up. Count the students standing. Count the students sitting. Have students use their TI-15s to find the sum of those two numbers.

2. Using their TI-15s, have students enter the number of students standing as the numerator, and the total number of students in the class as the denominator.

3. Have students press [Simp] and [Enter] to simplify the fraction. Tell students that this fraction represents the students in the class who raised their right hands.

4. Ask students if the hand each student raised was random.

5. Although students may or may not believe the event was random, point out that most right-handed students probably raised their right hands, and most left-handed students probably raised their left hands.

6. Given this information and the fraction calculated, have the students decide if the probability of randomly choosing a right-handed student from the class is greater or smaller than the probability of choosing a left-handed student.

In the Long Run (cont.)

Explaining the Concept

1. Ask the class: *What are some random events?* Students should suggest flipping coins, tossing number cubes, spinning spinners. Tell them that they are going to investigate flipping coins. Ask them what you call the two sides of a coin. Ask them what they expect to happen if you flip a coin 10 times.

2. Show students how to create a tally sheet by making marks for each event on the sheet. Explain that in a tally, you make a mark for each outcome in the category. After four marks have been made, the fifth one is marked by crossing the first four. Ask students why this is a good idea.

3. Explain that the frequency is the number of times each outcome occurred. Ask students to find the frequency for heads and tails. Explain that each flip of the coin is called a *trial*. Ask students how they can find the total number of trials. They may say counting the marks or adding the frequency.

4. Ask students what part of the flips were heads. Show them how to enter the fraction $\frac{6}{10}$ into their TI-15s by pressing [6] [n]. Next, press [1] [0] [d] to enter the denominator. Show them that they can simplify the fraction by pressing [Simp] and [Enter].

5. Ask students if they think the results of the coin flip shown above are realistic. Explain that $\frac{1}{2}$ is the *theoretical probability* that a coin will land heads up. To help students understand the word *theoretical*, ask them what a theory is. Tell them that they are going to do an experiment to find out what really happens.

6. Divide the class into groups of three or four. Distribute copies of the *Heads or Tails* activity sheet (pages 205–206; page205.pdf). Give each student a coin and a paper cup. Explain that they are going to shake the coin in the cup and then dump it on the table rather than actually flipping the coin.

7. Have students complete items 1–9 on their *Heads or Tails* activity sheets. Ask several students to report their results. Tell them that the fractions they found are *experimental probabilities* because they are the results of an experiment. Ask them how the experiment might be improved. They should suggest more trials. Ask them to compile the group information for the class chart and report the results to you. Write the frequencies for each group on the board or overhead. Have students find the class totals. Then have students complete items 10–15 on their activity sheets.

Lesson 21

In the Long Run (cont.)

Applying the Concept

1. Distribute copies of the *Fair Play* activity sheet (page 207; page207.pdf) so students can complete them. Each student should have a number cube and a paper cup. They will shake the number cube in the cup. Remind students that each group should report its results to you as they complete the second chart.

2. Ask students to explain what they have learned about random events. Have them compare experimental probability to theoretical probability. They should realize that for any single random event, it is impossible to predict the next result. They should understand that in the long run, experimental probability gets close to theoretical probability.

Differentiation

- **Below Grade Level—** Have students work in pairs, with one tossing the coin or number cube and the other recording the tally sheet. Have them take turns entering the numbers on the TI-15.

- **Above Grade Level—** Ask students to consider what would happen if they considered even and odd numbers on the number cubes. Then, ask them to consider numbers that are divisible by three.

Extending the Concept

Have students plan another probability experiment and record the results. They could use number cubes, spinners, or playing cards. Students should predict the theoretical probability and then find the experimental probability based on the results of their experiment.

Name _____

Heads or Tails

Directions: Answer the questions below.

1. If you flipped a coin twenty times, how many times do you think it would land heads up? How many times do you think it would land tails up?

   _____

2. Write a fraction that shows what part of the time you predict the coin will land heads up.

   _____

3. Enter the fraction on your TI-15 by entering the numerator and then pressing [n]. Next, enter the denominator and press [d]. Simplify the fraction by pressing [Simp] and [Enter]. Continue to press it until the answer stays the same. What is the simplified fraction?

   _____

4. Write a fraction that shows what part of the time you predict the coin will land tails up.

   _____

5. Like in question 3 above, use your TI-15 to find the fraction in simplest terms. What is the fraction?

   _____

6. Flip a coin 20 times. Use the tally sheet below to record your results.

Coin Face	Tally	Frequency
Heads		
Tails		

7. You have found the *experimental probability* of tossing heads or tails. The number of times you flipped the coin is called the *number of trials*. What was your number of trials?

   _____

8. Were the results the same as your prediction? _____

Heads or Tails (cont.)

Directions: Answer the questions below.

9. The fraction $\frac{10}{20}$, has a 10 (half the flips) in the numerator and a 20 (the total flips) in the denominator. The simplified fraction is called the *theoretical probability*. What is the theoretical probability of a coin landing heads up?

_____

Directions: Ask each of the students in your group about their results. Complete the chart below for each of the students in your group. Include your results, as well.

Name	Number of Heads	Simplified Fraction	Number of Tails	Simplified Fraction
Total				
Class Total				

10. Were anyone's results the same as the theoretical probability? _____

11. What was the experimental probability for your group? _____

12. What was the number of trials? _____

13. Report your group's results to your teacher. After the class totals have been posted, add them to the chart above. Find the simplified fraction for the class.

14. How many trials does the class total represent? _____

15. Look at your results, your group's results, and the class results. Compare the experimental probability with the theoretical probability. Which experimental probability was closest to $\frac{1}{2}$ or 0.5? Why do you think this is?

_____

_____

Name _____

Fair Play

Lesson 21

Directions: Answer the questions below.

1. How many sides does a cube have? _____

2. If you tossed the number cube 60 times, how many times would you expect to get a 3?
   _____

3. Toss a number cube 60 times and record the results on the chart below. Use your TI-15 to simplify the fraction.

Number on Cube	Tally	Number	Fraction	Simplified Fraction	Experimental Probability
1					
2					
3					
4					
5					
6					

4. Use the table to record the total results for your group and then the class.

Number on Cube	Group	Group Fraction	Group Experimental Probability	Class	Class Fraction	Class Experimental Probability
1						
2						
3						
4						
5						
6						

5. How many trials did you complete? _____

6. How many total trials did your group complete? _____

7. How many total trials did your class complete? _____

8. Which set of results was closest to the theoretical probability? _____

Lesson 22

Bowling for Dollars

Overview

Students will learn to find averages and investigate the applications of average.

Mathematics Objective

Understands the concept of the mean.

TI-15 Functions

- Constant feature
- Operations
- Repeated operations

Materials

- TI-15s
- *Rolling Along* activity sheet (pages 211–212; page211.pdf)
- *On a Roll* activity sheet (page 213; page213.pdf)

 Vocabulary

Complete the *Frayer Model* (page 20) vocabulary activity using the words below. Definitions for these words are included on the Teacher Resource CD (glossary.pdf).

- average (mean)
- quotient
- sum

Warm-up Activity

1. Divide the class into four groups. Write the groups of numbers in steps 2–5 on the board for each group. Have each group use the TI-15 to add its numbers and then divide by 5. Have them record the sum and the quotient.

2. Have the first group add 28, 29, 30, 31, and 32, and then divide by 5.

3. Have the second group add 24, 29, 30, 31, and 36, and then divide by 5.

4. Have the third group add 21, 23, 29, 33, and 44, and then divide by 5.

5. Have the fourth group add 10, 24, 30, 35, and 51, and then divide by 5.

6. Ask the groups their sums and quotients. Ask them why they are all the same.

Bowling for Dollars (cont.)

Explaining the Concept

1 Explain to students that they will need to raise money to pay for a field trip. Ask students if they have any ideas on how to raise money. List their ideas on the board or overhead. After hearing their suggestions, recommend a fund-raiser in the form of a bowling tournament.

2 Ask the class: *Have you ever gone bowling? What was your highest score? What is the score of a perfect game in bowling? What do you think is an average score for a fourth grader?* A perfect score in bowling is 300, but it is very difficult to achieve, requiring the bowler to roll a strike every time. A good score for most fourth graders would be about 100. Tell the students that they are going to look at the scores for a bowling tournament.

3 Remind students that you compute an average by adding all the scores and dividing by the number of scores. Ask them to explain the results of the warm-up activity in terms of averages.

4 Remind students how to use the constant feature (Op1) to easily repeat a computation. Show them that pressing [Op1] [÷] [3] [Op1] will allow them to divide any number by three by just entering the number and pressing [Op1]. Have them practice dividing several numbers by three.

5 Distribute copies of the *Rolling Along* activity sheet (pages 211–212; page211.pdf) to students. Work through the average for Julia on the red team to be sure that students understand that they should find the sum of the three numbers, and then divide it by three. Have them complete their charts.

6 After students have completed the charts, ask members of the class to share their answers. Be sure that all students have correct answers before proceeding.

7 Discuss how to find the team averages. Some students may wish to divide the total scores by 12 because they represent a total of 12 games. Other students may wish to average the average scores for each team member. Be sure to point out that the repeated average on the red team must be entered twice, once for each player who got that average.

8 Ask students if they can tell which team won without finding the average. They should realize that the total points for the team indicates the winner because the same number of games were bowled by each team.

9 Allow students to complete the remainder of the activity sheet in groups. After students have completed the activity sheet, ask them how they could figure out the score that Tamryn needed in number 8 on their activity sheets.

Lesson 22

Bowling for Dollars (cont.)

Applying the Concept

1 Distribute copies of the *On a Roll* activity sheet (page 213; page213.pdf). When students have completed the activity sheets, be sure that they understand that you can always find the total from the average as long as you know how many numbers are used. You may need to help students understand how to find a missing score when they know the final average.

2 Ask students to explain how they found each student's score on number 1. Then ask them how they figured out the scores for Mr. Garcia's class in Game 2.

3 Tell students that Emily bowled five games and had scores of 125, 178, 202, and 175 for four of them. Her average was 196. Ask them how to find her fifth score. (*It is 300—a perfect game!*)

Differentiation

- **Below Grade Level—** Help students visualize an average by finding the average of 25 and 75, and then finding the average of 40, 25, and 55. Then, have them find the amount above or below the average for each number. They should see that the total above is equal to the total below. (e.g., The average of 40, 25, and 55 is 40. The difference between 40 and 25 is the same as the difference between 55 and 40.)

- **Above Grade Level—** Have students explore *weighted average* and when it is used. Have them create a Venn diagram to compare *average* and *weighted average*.

Extending the Concept

Have students discuss how average is not always a good indication of the middle of a group. Suggest they think of the average height of 5 people if all them are fourth graders. Find the average height of five students in the class. Then, have students find the average height of five adults.

Name _____

Rolling Along

Directions: Read the text, follow the instructions, and answer the questions.

Your class is participating in a charity event at a bowling alley. The class is sending three teams of four bowlers each. Each student will bowl three games. The team with the highest average will earn a number of dollars equal to their average to give to a charity. The scores for the Red Team and the Blue Team are shown in the chart below.

Red Team	Game 1	Game 2	Game 3	Total	Average
Julia	104	98	113		
Jorge	128	76	102		
Nadia	102	102	102		
Jeffrey	156	144	150		
Blue Team	**Game 1**	**Game 2**	**Game 3**	**Total**	**Average**
Barry	58	145	121		
Serena	132	125	151		
Raymond	44	57	70		
Anna	97	108	101		

1. Use your TI-15 to find the total number of points scored by each student and write them in the chart.

2. Use your TI-15 to find the average for each student. Because you are going to divide by 3 for each student, you can use the [Opl] key. First, press [Opl] [÷] [3] [Opl]. Then, enter the total of the three games and press [Opl]. Find the averages for each of the students and write them in the chart.

3. Which three students had the same averages? What is true about their total points?

   _____

   _____

4. What is each team's average? How did you find the averages?

   _____

   _____

Rolling Along (cont.)

Directions: Read the text, follow the instructions, and answer the questions.

The Purple Team started later than the Red and Blue teams. All the members on the Purple Team had finished two games when the Red Team and Blue Team finished all three games. Their scores are on the chart below. They know the results from the other teams and want to know how they can beat the other teams.

Purple Team	Game 1	Game 2	Total after Game 2	Game 3	Total	Average
Joe	86	105				
Tamryn	128	129				
Mika	100	150				
Eugene	88	120				

5. What is the lowest number of points the Purple Team needs to score to beat the Red Team? What would their team average be?

6. What is the lowest number of points the Purple Team needs to score to beat the Blue Team? What would their team average be?

7. What is the lowest score Tamryn could get on her third game to beat Jeffrey?

8. Assume that Tamryn beats Jeffrey by 1 point. Write the score for her third game in the chart. How many total points do the other team members need to beat the Red team?

9. Give scores to Joe, Mika, and Eugene so that the Purple Team wins by the lowest possible amount. Fill in the chart above.

10. If the Purple Team won by the lowest possible score, how much money was donated to charity?

Name _____

On a Roll

Directions: Read the text, follow the instructions, and answer the questions.

In a bowling tournament, Mrs. Howard's class is going to bowl against Mr. Garcia's class. Each class chooses its best bowlers for a team of four. Each student will bowl two games.

1. In the first game, the students on Mrs. Howard's team all bowled the same score. The team scored a total of 572 points in that game. What was the score for each student on the team? Fill in those points in the chart below.

   _____

2. In the first game, the students on Mr. Garcia's team averaged 141. How many total points did they score?

   _____

3. On Mr. Garcia's team, the scores for Faraah, Ajay, and Gracie were 147, 138, and 150, respectively. What was Ziva's score?

   _____

Mrs. Howard's Team	Game 1	Game 2	Average	Mr. Garcia's Team	Game 1	Game 2	Average
Evelyn			138	Faraah			129
Hans			150	Ajay			141
Bianca			122	Gracie			158
Ernest			146	Ziva			110

4. Find the scores for each student in Game 2, and write them in the chart.

5. What was the average score for Mrs. Howard's team? _____

6. What was the average score for Mr. Garcia's team? _____

7. Which team won the bowling tournament?

   _____

Lesson 23

It's in the Bag

Overview

Students will learn to estimate the compositions of samples through repeated trials.

Mathematics Objective

Understands equivalent forms of basic fractions and decimals and when one form of a number might be more useful than another.

TI-15 Function

- Fractions

Materials

- TI-15s
- chips in three different colors
- small lunch bags
- *Bag of Chips* activity sheet (pages 217–218; page217.pdf)
- *Mystery Chips* activity sheet (page 219; page219.pdf)

Vocabulary

Complete the *Music Makers* (page 18) vocabulary activity using the words below. Definitions for these words are included on the Teacher Resource CD (glossary.pdf).

- fraction
- sample
- trial
- frequency
- tally marks

 Warm-up Activity

1. Tell students that $\frac{7}{10}$ of people prefer dogs to cats. Ask them how many people this means.

2. Ask students if every 7 out of 10 people prefer dogs to cats, how many people would prefer dogs if there were 100 people surveyed.

3. Ask them how many people would prefer dogs if 250 people were asked. Show them how to find $\frac{7}{10}$ of 250 on their TI-15s by pressing 7 n 1 0 d × 2 5 0 Enter.

4. Have students calculate how many people would prefer dogs if 850 people were asked. Have them follow the procedure from step 3 using the number of 850 instead of 250.

#50616—30 Mathematics Lessons Using the TI-15 © Shell Education

It's in the Bag (cont.)

Explaining the Concept

1 Tell students that there is a bag of with 10 chips in it, $\frac{1}{10}$ of the chips are red, $\frac{3}{10}$ chips are blue, and $\frac{6}{10}$ chips are yellow. Ask students how many of each color there would be if there were 100 chips in the bag. Ask them how many of each color there would be if there were 50 chips in the bag. Have them use their TI-15s to find how many of each color there would be if there were 80 chips in the bag. To do this, students should use the same process they followed during the warm-up.

2 Tell students that there is another bag with 20 chips in it. The bag has 12 red chips, 6 blue chips, and 2 yellow chips. Ask students to write fractions that represent the parts of the chips that are red, blue, and yellow. Ask them how they determined the denominator of the fraction. Show them how to enter the fractions by entering the numerator, pressing the [n] key, entering the denominator, and pressing the [d] key.

3 Tell the class that they are going to learn to analyze the contents of a bag of chips by drawing out one chip at a time and replacing it before drawing another chip. There may be two or three colors of chips. Show students a bag containing 10 chips—3 red, 2 blue, and 5 yellow. Ask them to find the fractions that represent each color. Ask students what they think would happen if they drew and replaced the chips 10 times.

4 Organize students into pairs and distribute copies of the *Bag of Chips* activity sheet (pages 217–218; page217.pdf). Give one bag of chips to each pair of students. Check that the bags all contain 10 chips with the same number of three different-colored chips. Tell students that all of their bags contain the same number of chips of each color. Have students work through the first three charts on the activity sheet.

5 After students have completed the third chart, have them report the results of the 20 trials to you. Write the results for each pair on the board or overhead. Ask students how many of them got the same results both times. Check to see if all the students have found the same results after combining the results for both partners. They may or may not. Find the totals for each color for the entire class. Have students enter the data on the chart and complete the activity sheet.

6 Ask students if they know for sure how many chips are in the bag. Have them look at the chips in the bag. The fractions they found were for a total of 10 chips. Ask them how they could get the same fractions with 20 chips, 30 chips, or even 50 chips.

© Shell Education   #50616—30 Mathematics Lessons Using the TI-15   **215**

Lesson 23

It's in the Bag (cont.)

Applying the Concept

1 Distribute copies of the *Mystery Chips* activity sheet (page 219; page219.pdf) to students. Each student will need a bag and a variety of chips. Tell students that they are going to put 10 chips of three colors in a bag and trade bags with partners. Each partner will try to guess the number of each color of chips in the bag. Explain the scoring rules as outlined on the activity sheet.

2 Students may draw and replace as many chips as they feel necessary, but remind students that if the game ends in a tie, the student with the fewest number of tally marks will win the game.

Differentiation

- **Below Grade Level**—Limit the number of colors of chips to two colors. Make one pile of 10 using two colors of chips. Make another pile of 20 chips using the same two colors. Make sure that both piles contain the same ratio of each color. Have students divide the group of 20 into two groups of 10, making sure to divide each color equally. Have students calculate fractions for each color in each group. Have them explain why the fractions for each group are the same.

- **Above Grade Level**—Have students investigate how random sampling is used.

Extending the Concept

Have students set up a tournament using bags of chips. Have them use varying numbers of chips and colors of chips at each round of the tournament to increase difficulty at each level. Have them show their tallies and fractions on the board at the upper levels so that the class can follow the progress of the tournament.

Name _____

Bag of Chips

Directions: Read the text, follow the instructions, and answer the questions.

You and your partner have a bag with 10 chips in it. Your task is to determine what the colors of the chips in the bag are and what part of the total number of chips each color is. One partner will draw out one chip at a time, and then replace it. The other partner will record your results with tally marks. You may NOT look into the bag.

Draw and replace chips from the bag 10 times. Record your results below.

Color	Tally	Frequency	Fraction

Now trade jobs with your partner and do it again. Enter the results on the chart below.

Color	Tally	Frequency	Fraction

1. Did you get the same results both times? Why do you think this is true?

   _____

2. How many of each color chips do you think are in the bag? Why do you think that?

   _____

   _____

Bag of Chips (cont.)

Directions: Read the text, follow the instructions, and answer the questions.

3. Combine the results from the first two charts on the chart below. How many trials does this represent? _____

Color	Frequency	Fraction

4. Are you positive that these are the correct fractions for the chips in your bag? _____

5. Everyone in the class has bags with the same numbers and colors of chips. Report your results to your teacher. Record the total class results on the chart below.

Color	Frequency	Fraction

6. Which set of results do you think are the most accurate (those from one student, two students, or the entire class)?

   _____

7. Take all of the chips out of your bag and count the colors. How many chips are there? Are the class fractions correct? If not, what are the correct fractions?

   _____

8. Suppose there were 2 bags of chips. How many chips would there be? How many of each color would there be?

   _____

9. Suppose there were 5 bags of chips. How many chips would there be? How many of each color would there be?

   _____

Name _____

Mystery Chips

Directions: Read the text, follow the instructions, and answer the questions.

You and your partner are going to play a game. You are each going to put 10 chips in a bag and trade bags. After drawing a chip out of the bag, record it in the chart and place it back in the bag. Keep drawing chips until you think you know how many of each color are inside the bag. Then, figure out the fractions for each color. When you correctly give the fraction of each kind of chip in the bag, you score a point. You only get two guesses for each bag of chips. You will fill and trade the bags four times. The person with the most points wins. If there is a tie, the person with the fewest number of tally marks wins.

Put 10 chips in a bag. Write down how many of each color you have used. Do not show your numbers to your partner. Trade bags with your partner. Use the charts below to keep track of the number of draws you have made. Keep track of the points next to your name. Change the number of each color in the bag for the next exchange.

Bag 1

Color	Tally	Frequency	Fraction

Bag 2

Color	Tally	Frequency	Fraction

Bag 3

Color	Tally	Frequency	Fraction

Bag 4

Color	Tally	Frequency	Fraction

Lesson 24

They Add Up

Overview

Students will find the experimental probability of rolling each possible sum on pairs of number cubes, make line graphs of their results, and find the theoretical probability of each sum.

Mathematics Objective

Determines probability using simulations or experiments.

TI-15 Functions

- Fraction
- Fraction to percent

Materials

- TI-15s
- pair of number cubes for each student (2 different colors in each pair)
- *Let's Get Rolling* activity sheet (pages 223–224; page223.pdf)
- *Toss Those Cubes!* activity sheet (page 225; page225.pdf)

 Vocabulary

Complete the *Sentence Frames* (page 16) vocabulary activity using the words below. Definitions for these words are included on the Teacher Resource CD (glossary.pdf).

- experimental probability
- frequency
- outcome
- percent
- prediction
- probability
- random
- sum
- tally sheet
- theoretical probability
- trial

Warm-up Activity

1. Give each student a pair of number cubes.

2. Ask each student to roll the pair of cubes. On the board or overhead, use tally marks to record the number of times each sum occurs on a chart like the one below. Begin by showing just the *Sum* and *Tally* rows on the chart. The other rows will be uncovered as the lesson continues. Ask students what the smallest sum is and fill it in the top row. Then, ask for the next largest and continue until all possible sums are listed. The sums 2 through 12 should occur. Ask students how many got each total. Make tally marks for each sum.

3. Have all students guess what the sum of the cubes on another roll will be and then have them roll again. Ask how many of their predictions were correct. Record these results and find the total number of times each sum occurred on the two rolls. Fill in the *Total* row on the chart.

Sum	2	3	4	5	6	7	8	9	10	11	12
Tally											
Total											
Fraction											
Percent											

Lesson 24

They Add Up (cont.)

 ## Explaining the Concept

1 Ask students how many rolls of number cubes are represented on the chart from the warm-up activity. Remind them that each roll of number cubes is called a *trial*. Ask them to write a fraction for each sum, showing what part of the rolls produced each sum. The numerator of each fraction should be the number of times each sum occurred. The denominator is the total number of trials.

2 Remind students how to enter a fraction into the TI-15 by entering the numerator and pressing [n]. Next, enter the denominator and press [d]. Show them that they can simplify the fraction by pressing [Simp] and [Enter].

3 Tell students that the results would be easier to interpret if they were expressed as a percent. Review with students the meaning of percent. Remind them that they can change a fraction on the TI-15 to a percent by pressing [▶%] and [Enter]. Have each student find the percents for the data on the chart and add their answers to the *Percent* row. Ask them to round the percents to the nearest whole percent.

4 Distribute copies of the *Let's Get Rolling* activity sheet (pages 223–224; page223.pdf). Tell students that they are going to conduct more trials and determine if the sums are random.

5 After students complete number 6 on the activity sheet, they will report the total number of times each sum occurred in their 50 trials. Find the total number of times each total occurred for the class and add it to the chart below. (You could assign students to be responsible for each sum.)

Sum	2	3	4	5	6	7	8	9	10	11	12
Total											

6 The total number of trials and the sum of all the totals should equal 50 times the number of students in the class. Analyze the class data together.

7 After students have completed number 10, show them how to place each point on the graph by finding the sum on the horizontal axis and moving up so that the point is across from the percent on the vertical axis.

8 After students have completed their graphs, discuss that the graph shows the *experimental probability* of each sum. Remind them that it is experimental because they found it by conducting a real-life experiment. Ask them how the graph is useful in analyzing their results. They should notice that it makes the differences in the percentages obvious with just a glance.

They Add Up (cont.)

Applying the Concept

1 Distribute copies of the *Toss Those Cubes!* activity sheet (page 225; page225.pdf). Tell them that they are going to find the *theoretical probability* of rolling each sum on a pair of number cubes. Remind them that they are going to need their copies of their *Let's Get Rolling* activity sheets.

2 When students have completed the sheet and the graph on their *Let's Get Rolling* activity sheets, discuss how the two graphs are similar and different. They should be approximately the same. Ask them why some of the sums are more likely than others. Discuss that although the results on each number are random, there are more combinations for each sum.

Differentiation

- **Below Grade Level—** Show students some of the combinations on the *Toss Those Cubes!* activity sheet using number cubes. Be sure that they understand that they need to consider the combination of 1 on the first cube and 3 on the second cube as different from 3 on the first cube and 1 on the second.

- **Above Grade Level—** Have students consider the possibilities if 3 cubes were tossed.

Extending the Concept

Have students consider 100, 200, 500, and 1,000 rolls and predict the number of times each sum will occur.

Name _____

Let's Get Rolling

Directions: Follow the steps below.

1. Roll your pair of number cubes 50 times. Make a tally mark on the chart below for each time the sum occurs. Then write the total for each sum in the *Total* row.

Sum	2	3	4	5	6	7	8	9	10	11	12
Tally											
Total											
Fraction											
Percent											

2. For each sum, write a fraction for the part of the total rolls in the *Fraction* row above.

3. Use your TI-15 to change each fraction to a percent.

4. Which sum(s) occurred most often? _____

5. Which sum(s) occurred the least often? _____

6. Which sums(s) have the highest experimental probability?
   _____

7. Fill in the class totals for each sum on the chart below What is the total number of trials for the class? _____

Sum	2	3	4	5	6	7	8	9	10	11	12
Total											
Fraction											
Percent											

Let's Get Rolling (cont.)

Directions: Use your findings from the previous page to answer the questions below.

8. Which sum(s) occurred most often? _____

9. Which sum(s) occurred the least often? _____

10. How did the class results compare to your results?
    _____

11. Graph the class results on the grid below by placing a point above each sum across from the percent. Connect the points with line segments.

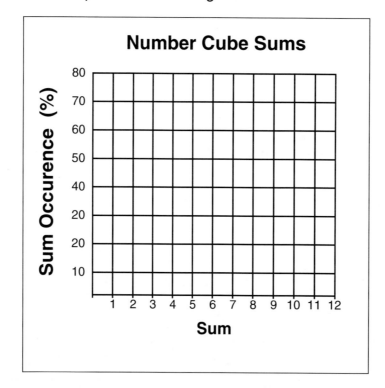

12. Which sums have about the same experimental probability? _____

13. Which sum(s) have the greatest experimental probability? _____

14. Which sum(s) have the smallest experimental probability? _____

15. How does the graph make it easy to answer questions 12–14? _____

16. If you were going to guess a sum on a roll of two number cubes, what would you guess?
    _____

Name _____

Toss Those Cubes!

Directions: Complete the charts, and answer the questions below.

1. This chart starts to list all possible tosses of two number cubes. Complete the chart.

1st Cube	2nd Cube	2nd Cube	2nd Cube	2nd Cube	2nd Cube	2nd Cube
1	1	2	3			
2	2	3	4			
3	3	4	5			
4	4	5	6			
5	5	6	1			
6	6	1	2			

2. How many different combinations are there? _____

3. List all of the combinations that equal each sum. Then, find the simplest fraction and percent that represent what part of the combinations give each sum.

Sum	2	3	4	5	6	7	8	9	10	11	12
Combinations											
Fraction											
Percent											

4. How do these results compare to the class results on the *Let's Get Rolling?* Make a broken line graph on the grid on the *Let's Get Rolling* activity sheet using this data. Use a different color, so you can tell the difference.

Lesson 25

Typically

Overview

Students will investigate mean, median, and mode and discover when each yields the best measure of central tendency.

Mathematics Objective

Understands that a summary of data should include where the middle is and how much spread there is around it.

TI-15 Functions

- Memory
- Memory recall
- Operations

Materials

- TI-15s
- *Birthday Weather* activity sheet (pages 229–230; page229.pdf)
- *Birthday Game* activity sheet (page 231; page231.pdf)

 ## Vocabulary

Complete the *Which Statement Is Accurate?* (page 17) vocabulary activity using the words below. Definitions for these words are included on the Teacher Resource CD (glossary.pdf).

- arithmetic mean
- average
- difference
- mean
- measure of central tendency
- median
- mode
- sum

Warm-up Activity

1. Have students use their TI-15s to find the average of 5, 10, 15, 20, 25, and 30.

2. Then, have them find the average of 1, 2, 3, 4, 5, and 90.

3. Ask students why the two lists have the same average.

4. Ask them if they think the average of the second list accurately describes the middle of the list.

Typically (cont.)

Explaining the Concept

1. Ask students how to find the average of the value of the coins in the bag that contains 10 pennies, 1 dime, and 5 quarters. When they find the average (*about 9 cents*), ask them if that means the typical coin in the bag is a dime. The students should agree that the typical coin is a penny because there are a lot more pennies than dimes or quarters in the bag.

2. Have students consider a group of adults of varying heights, including the shortest person in the world and the tallest person in the world. Imagine that they could line them up from shortest to tallest. Ask them if the person in the middle of the line would be considered typical.

3. Explain to students that they have been looking at *measures of central tendency*. Measures of central tendency include the mean, the median, and the mode. The *mean* is the same as the average, the *median* is the middle number when all the numbers are listed from least to greatest, and the *mode* is the number that occurs most frequently. Have students identify how they used the mean, median, and mode in the discussion of the coins and height.

4. Distribute copies of the *Birthday Weather* activity sheet (pages 229–230; page229.pdf). Ask students what kinds of party plans might depend on the weather.

5. After students have completed questions 1–8, ask them why the average was not halfway between the lowest and highest temperatures. They should discuss how the extremes might not be representative of the rest of the temperatures. Remind students that the average is also called the *mean*.

6. Show students how to find the middle of a list that has an even number of items. Have students put the data in numerical order. Then, show them how to count from both ends to find the two middle numbers. Students must then add the two middle numbers together and divide by two to find the median.

7. Ask students to complete items 9–18. When they have finished, discuss the median and mode. Ask them if they think the median of this list might be considered typical of the temperature on February 15 in Columbus, Ohio.

8. Have students complete the activity sheet. When they have finished, discuss the three measures of central tendency that they have found. Ask them how each is a good or poor indicator of the typical day.

Typically (cont.)

Applying the Concept

1 Distribute copies of the *Birthday Game* activity sheet (page 231; page231.pdf). Point out that many of the guests' heights listed on the activity sheet contain 4 feet and some inches left over. An easy way to compare the heights would be to convert all the measurements into inches. Remind students that 1 foot = 12 inches.

2 Show students how to store a number in the TI-15 memory by using the following example: [4] [×] [1] [2] [Enter] [▶M] [Enter]. They should see the equation with the solution 48, and an M at the top of the display. Have them clear the numbers from the display and then press [MR/MC] to recall the solution. Explain that the number will remain in the memory until they press [MR/MC] twice in a row. Point out that if they press [MR/MC] after clearing the memory, the display will show 0.

Differentiation

- **Below Grade Level—** Use groups of students as visual representations of the measures of central tendency. Have an odd number of students line up in order of height and identify the median. A group of more girls than boys could also demonstrate the concept of mode.

- **Above Grade Level—** Ask the students to compile a list that has the same mean, median, and mode. Then, ask them to find a list of 10 numbers with an average of 35, a median of 40, and a mode of 27.

Extending the Concept

- Discuss if it makes sense to have a mode if all the items in a list are different or if there is no one item that occurs most frequently. Explain that these lists have no mode.

- Illustrate the mode as a good measure for a group of items by looking at a group of straws where most of the straws are of one length, with a few shorter straws. Also ask the students to find the mode of the coins used earlier in the lesson.

Name _____

Birthday Weather

Directions: Read the text and answer the questions.

Adrienne's birthday is February 15. She is planning a big party. The activities she might plan depend on the weather. From 1987 to 2009, the high temperatures (in °F) on February 15 for her city are listed below.

28°, 7°, 37°, 59°, 16°, 59°, 32°, 45°, 45°, 29°, 30°, 45°, 50°, 41°, 42°, 52°, 30°, 28°, 64°, 57°, 17°, 35°, and 34°

Adrienne wants to determine the high temperature for a typical day on February 15. She thought that one way would be to find the average temperature.

1. How many temperatures are listed? _____

2. Explain why the number of temperatures does not equal 2009–1987.

   _____

3. To the nearest tenth, what is the mean temperature for February 15 in Adrienne's city?

   _____

4. What is the highest temperature in the list? _____

5. What is the lowest temperature in the list? _____

6. What is the difference between the mean and the highest temperature in the list?

   _____

7. What is the difference between the mean and the lowest temperature in the list?

   _____

8. Why isn't the mean of the temperatures exactly halfway between the low and high?

   _____

© Shell Education   #50616—30 Mathematics Lessons Using the TI-15   **229**

Birthday Weather (cont.)

Directions: Read the text and answer the questions.

Adrienne asked her teacher, Mr. Jenkins, if there were any other ways to think of the typical high temperature besides using the mean. Mr. Jenkins told her to consider the *median*.

9. Do you think the median will be smaller or larger than the mean? Why?

_____

10. To find the median, put the temperatures in order from smallest to largest.

_____

_____

11. Does your list have a middle number? What is it? _____

12. What is the difference between the largest number in the list and the median? _____

13. What is the difference between the smallest number in the list and the median? _____

14. Are your answers to items 12 and 13 the same? Why?

_____

15. Is the median larger or smaller than the mean? Why?

_____

Mr. Jenkins told Adrienne that there is one more way to measure central tendency, the *mode*.

16. Is there a mode on the list? What is it? _____

17. What do you think the high temperature would be on a typical February 15 in Adrienne's city? Do you think she'll be able to have her party outside?

_____

18. Which measure of central tendency did you use to answer question 17? Why did you choose it?

_____

_____

Name _____

Birthday Game

Directions: Read the text and answer the questions.

The heights of the guests at Adrienne's birthday party are listed below. Adrienne is going to organize the data and award prizes to the guest whose height is closest to the mean and to the guest whose height is the median of the party.

4 ft. 2 in.		4 ft. 9 in.		5 ft. 3 in.		4 ft. 9 in.	
4 ft. 9 in.		4 ft. 5 in.		4 ft. 6 in.		4 ft. 7 in.	
4 ft. 10 in.		4 ft. 2 in.		4 ft. 7 in.		4 ft. 11 in.	
4 ft. 9 in.		4 ft. 3 in.		4 ft. 0 in.		4 ft. 11 in.	

1. Why would it be difficult to find the mean of their heights?

   _____

2. How could you change their heights into inches?

   _____

3. Use your TI-15 to find the number of inches in 4 feet. _____

   Store the answer to question 3 in the memory by pressing ▶M Enter. To use the stored value in finding the height of the first height in the chart, press MR/MC once, and then + 2 and Enter.

4. Now, find the heights in inches. Record that information in the chart above. Then rewrite the list from smallest to largest below.

   _____

5. What is the mean of the party guests? _____

6. What is the median of the party guests? _____

7. What is the mode of the party guests? _____

8. Which measure of central tendency do you think best describes the party guests? Explain your answer.

   _____

   _____

© Shell Education #50616—30 Mathematics Lessons Using the TI-15

Power Lesson 5

Graph It!

Overview

Students will learn to use and make circle and bar graphs to represent data.

Mathematics Objective

Organizes and displays data in circle graphs, bar graphs, and double bar graphs.

TI-15 Functions

- Fractions to percent
- Operations

Materials

- TI-15s
- *Frozen Flavors* activity sheet (pages 237–239; page237pdf)
- *A Flavorful Dessert* activity sheet (page 240; page240.pdf)
- *Favorite Flavors* activity sheet (page 241; page241.pdf)
- *Top That!* activity sheet (page 242; page242.pdf)

Vocabulary

Complete the *Word Wizard* (page 21) vocabulary activity using the words below. Definitions for these words are included on the Teacher Resource CD (glossary.pdf).

- bar graph
- circle graph
- data
- double bar graph
- percent

Warm-up Activity

1. Divide students into two groups based on the dates they were born. Group one will consist of students born on an odd day. Group two will consist of students born on an even day.

2. Find the total number of students in the class. Then find the total number of students for each group. Ask each group to write a fraction representing the portion of students born on an even or odd day.

3. Have the groups enter the fraction on their TI-15s. Then have them change the fraction to a percent by entering the numerator, pressing [n], entering the denominator, pressing [d], and then pressing [▶%]. Have students round the answer to the nearest whole percent.

4. Discuss the percentages and ask students ways they could represent this information.

#50616—30 Mathematics Lessons Using the TI-15 © Shell Education

Graph It! (cont.)

Part One
Explaining the Concept

1 Ask students to think about their favorite frozen yogurt flavors. Tell them that the choices are vanilla, chocolate, strawberry, peanut butter, or cake batter. Have each the student come to the board and write the flavor they would chose. Allow students to write anywhere on the board. Ask students what the class would like. They should suggest that the information needs to be organized so that they can better analyze the data. Ask them how they could organize the data.

2 Recreate the chart below on the overhead or board and have students help fill it in. Ask students how this chart helps them keep track of the data accurately. Keep this information. Students will need this data to complete the *A Flavorful Dessert* activity sheet (page 240; page240.pdf).

Flavor	Number
vanilla	
chocolate	
strawberry	
peanut butter	
cake batter	

3 After it is completed, ask students if they can think of ways they could display the data in the chart. Make sure students mention circle graphs and bar graphs.

4 Distribute copies of the *Frozen Flavors* activity sheet (pages 237–239; page237.pdf). Explain to students that Mrs. Wheeler and Mr. Ferraro asked their students the same questions about frozen yogurt flavors. Mrs. Wheeler decided to display her class data in a circle graph. Draw the circle graph below on the board or overhead.

Favorite Frozen Yogurt in Mrs. Wheeler's Class

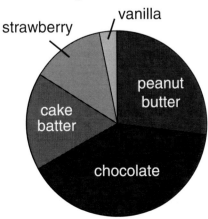

Graph It! (cont.)

Part One
Explaining the Concept (cont.)

5 Tell students that this type of graph is called a *circle graph*. Ask them why each category makes sense. Point out that the graph has a title and that each category is identified. Tell them that, in many cases, the segments are colored different colors. Explain that each part of the circle was constructed from the percent of the students who gave each response. Have them record their guesses for these percents on their activity sheets.

6 Now recreate the graph below on the board or overhead. Ask students how close their guesses were to the correct percents. Tell them that circle graphs typically show percents because they represent parts of a whole.

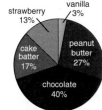

Favorite Frozen Yogurt in Mrs. Wheeler's Class

7 Ask students how they can find the number of students with each preference. Make sure students understand how to multiply decimals (.27 x 30) by solving the problem below on the board.

```
    30
  x .27
   210
   600
  8.10
```

8 Then remind students that the TI-15 can find 27% of 30 for them. Have them press the following keys: 2 7 % x 3 0 Enter. Remind students to round to the nearest whole number. Tell students this activity sheet will be completed later in the lesson.

Applying the Concept

1 Distribute copies of the *A Flavorful Dessert* activity sheet (page 240; page240.pdf). Explain to students they are going to create their own circle graphs. Have students copy the results from the class survey you took at the beginning of the lesson into the *Number* column on the chart. (See page 233.) Model for students how to divide the circle for the first entry.

2 Have students switch papers with partners to see if their circle graphs match. If time permits, have students add color to their newly-created circle graphs.

Graph It! (cont.)

Part Two
Explaining the Concept

1 Have students return to their copies of the *Frozen Flavors* activity sheet (pages 237–239; page237.pdf). Previously, students had studied Mrs. Wheeler's circle graph. Mr. Ferraro decided to display his class data in a bar graph. Draw the bar graph below on the board or overhead.

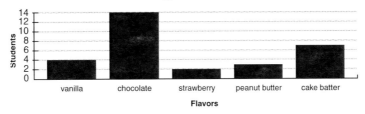

2 Have students look at the bar graph and describe how it was constructed. Discuss the scale on the vertical axis. Tell them that the numbers on the vertical axis must increase by the same amount for each line. For this bar graph, it increases by two. Work as a class to answer the questions regarding Mr. Ferraro's bar graph.

Applying the Concept

1 Distribute copies of the *Favorite Flavors* activity sheet (page 241; page241.pdf). Explain to students that they are going to figure out the number of people who voted for each flavor of frozen yogurt and create a bar graph to display the information.

2 Have students switch papers with partners to see if their bar graphs match. If time permits, have students add color to their newly-created bar graphs.

Part Three
Explaining the Concept

1 Have students return to their copies of the *Frozen Flavors* activity sheet (pages 237–239; page237.pdf). Remind students about Mrs. Wheeler's and Mr. Ferraro's plan to have a frozen yogurt party. Previously students had studied Mrs. Wheeler's circle graph and Mr. Ferraro's bar graph. They still need to figure out the top four flavors of frozen yogurt so they can order the machines for the party. The two teachers decided to make a double-bar graph.

2 Explain to students how a double-bar graph works. Have students look at the graph on page 239. Tell them that in this graph each class is represented by a different color, thereby breaking the data into two categories. The data is now easier to compare. Have students answer the questions regarding the double-bar graph to complete the activity sheet.

© Shell Education   #50616—30 Mathematics Lessons Using the TI-15   **235**

Graph It! (cont.)

Applying the Concept

1. Distribute copies of the *Top That!* activity sheet (page 242; page242.pdf). Conduct a class survey and have students fill out the *Number* column in the chart.

2. After students have completed their circle graphs and double-bar graph, discuss how the graphs help to show an immediate idea about the data. Be sure that the students see that a circle graph is especially useful for seeing the relative proportions of the categories, and a double-bar graph makes it easy to see how the categories are different.

Differentiation

- **Below Grade Level—** Provide students with manipulatives to use to represent the data being collected.

- **Above Grade Level—** Have students compare the same data using multiple types of graphs. Instruct students to choose which graph most accurately displays the data. Students should write at least two sentences explaining why the graph they chose is best.

Extending the Concept

Show students a broken line graph of data collected over time. One example might be a child's height over several years. The broken line graph makes it easy to see increases and decreases and the rate of increase or decrease over time. Have students collect data and create their own broken line graphs.

Name _____

Frozen Flavors

Directions: Read the text, study the graphs, and answer the questions.

Mrs. Wheeler and Mr. Ferraro's classes are going to have a frozen yogurt party. They decided to rent four frozen yogurt machines. The students in both classes voted on which frozen yogurt flavors they would like to have. The two teachers compiled the results and made graphs to show the class preferences.

Mrs. Wheeler made the circle graph below. Use this graph to answer the questions.

Favorite Frozen Yogurt Flavors in Mrs. Wheeler's Class

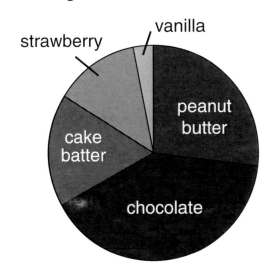

1. List the class preferences in order starting with the flavor that most students chose.

   _____

   _____

2. Estimate what percentage of the class prefers each flavor by filling in the chart below.

Favorite Frozen Yogurt Flavors in Mrs. Wheeler's Class

	Vanilla	Chocolate	Strawberry	Peanut Butter	Cake Batter
Percentage (%)					

Frozen Flavors (cont.)

Directions: Answer the questions below.

3. Mrs. Wheeler's students wanted more precise information, so she showed them the more informative graph below.

Favorite Frozen Yogurt Flavors in Mrs. Wheeler's Class

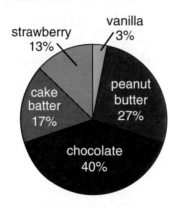

How close were your estimates? _____

4. There are 30 students in Mrs. Wheeler's class. How many students preferred each flavor? Fill out the chart below.

Favorite Frozen Yogurt Flavors in Mrs. Wheeler's Class

	Vanilla	Chocolate	Strawberry	Peanut Butter	Cake Batter
Number of Students					

Mr. Ferraro made a bar graph for his class. The graph is shown below.

Favorite Frozen Yogurt in Mr. Ferraro's Class

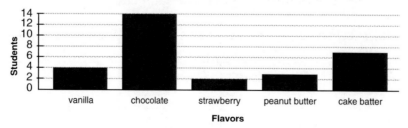

5. How many more students preferred chocolate than strawberry? _____

6. How many students are in Mr. Ferraro's class? _____

Frozen Flavors (cont.)

Directions: Read the text and anwer the questions below.

Mrs. Wheeler and Mr. Ferraro still needed to figure out which four frozen yogurt flavors their students preferred. They decided to make a double-bar graph to better view the data. The graph is shown below. Use the graph and your TI-15 to answer the questions.

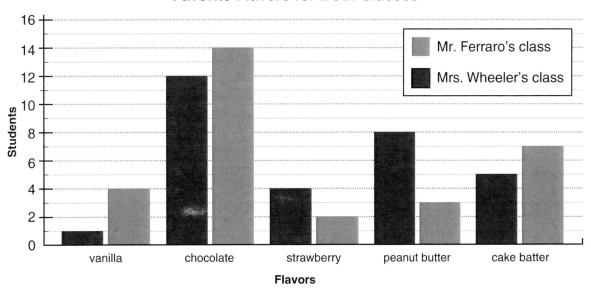

7. How many total students were surveyed? _____

8. How did you find the answer to question 7? _____

9. How many more students preferred cake batter in Mr. Ferraro's class than in Mrs. Wheeler's class? _____

10. Which class preferred vanilla over strawberry? _____

11. Which class preferred peanut butter over cake batter? _____

12. What four frozen yogurt flavors do the teachers need to order for the party?

13. If Mrs. Wheeler and Mr. Ferraro both voted for vanilla, would that change the outcome? Why or why not?

Power Lesson 5

Name _____

A Flavorful Dessert

Directions: Read the text, fill in the table, and answer the questions.

At the beginning of this lesson, your teacher conducted the same survey that Mrs. Wheeler and Mr. Ferraro used in their classes. Write in the results from that survey in the *Number* column of the chart below.

Flavor	Number	Fraction	Percent
Vanilla			
Chocolate			
Strawberry			
Peanut Butter			
Cake Batter			

1. How many students responded? _____

2. In the third column, write a fraction showing what part of the class preferred each flavor.

3. Enter each fraction into your TI-15 by entering the numerator, pressing the [n] key, entering the denominator, and pressing the [d] key. Then, change the fraction to a percent by pressing [▶%]. Enter the percent into the fourth column of the chart.

4. In the space below, use the information above to create a circle graph. Remember to estimate your sections carefully!

240 #50616—30 Mathematics Lessons Using the TI-15 © Shell Education

Name _____

Power Lesson 5

Favorite Flavors

Directions: Read the text, study the graph, and use the information to create a bar graph.

Frozen Yogurt Palace recently conducted a survey to find out which of their frozen yogurt flavors are the most popular. They surveyed 132 people. The survey results are shown on the circle graph below. Use the circle graph to create a bar graph. Remember, you will need to use your TI-15 to change the percentages to the number of people who chose each flavor. Round your answer to the nearest whole number. Fill in the table first and then use that data to create a bar graph in the space below.

Favorite Flavors at Yogurt Palace

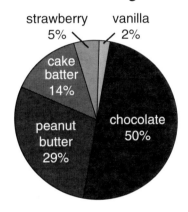

	Vanilla	Chocolate	Strawberry	Peanut Butter	Cake Batter
Number of People					

© Shell Education

#50616—30 Mathematics Lessons Using the TI-15

241

Name _____

Top That!

Directions: Read the text and answer the questions.

Conduct a survey of two classes to find which frozen yogurt toppings are the most popular. Fill in the results in the *Number* column below. Then use your TI-15 to complete the chart.

Class One

Topping	Number	Fraction	Percent
sprinkles			
gummy bears			
fruit			
hot fudge			
nuts			

Class Two

Topping	Number	Fraction	Percent
sprinkles			
gummy bears			
fruit			
hot fudge			
nuts			

Now it's time to organize and display your data. First, create a double-bar graph. When you have completed your graph, combine the data from the two charts above and use this new data to create a circle graph. Use both of your graphs to answer the questions below.

1. How many total students preferred gummy bears? _____

2. What total percentage of students preferred sprinkles? _____

3. Between the two classes, what are the three most popular toppings? _____
_____

4. Which graph was easier to use to answer the questions? Why?
_____
_____

Answer Key

Lesson 1: Close Enough
Round It Up (pages 45–46)

Week One			Week Two			Week Three	
Student	$ Total	Student	$ Total	Student	$ Total		
Raj	$3.72	Raj	$10.55	Raj	$6.19		
Jada	$8.82	Jada	$2.42	Jada	$2.03		
Estelle	$1.73	Estelle	$1.47	Estelle	$12.64		
Alex	$9.51	Alex	$3.28	Alex	$7.46		

Student	$ Total
Raj	$20.46
Jada	$13.27
Estelle	$15.84
Alex	$20.25

Method 1 (Rounded)
$20.00
$13.00
$16.00
$20.00

Student	Week 1	Week 2	Week 3	Method 2 (Rounded)
Raj	$4.00	$11.00	$6.00	$21.00
Jada	$9.00	$2.00	$2.00	$13.00
Estelle	$2.00	$1.00	$13.00	$16.00
Alex	$10.00	$3.00	$7.00	$20.00

1. No, the totals were not always the same for the two methods.
2. Raj had different totals for the two methods. He had different totals because the actual total is very close to 0.5 when the amounts were rounded using the second method.
3. Answers will vary.
4. Answers will vary.

More Money (page 47) (cont.)

Student	Week 1	Week 2	Week 3	Method 2 (Rounded)
Raj	$8.60	$22.30	$6.70	$37.60
Jada	$4.30	$3.70	$14.80	$22.80
Estelle	$2.60	$3.20	$17.70	$23.50
Alex	$4.20	$9.60	$2.90	$16.70

Raj and Alex had different totals for the two methods.

Week One		Week Two		Week Three	
Student	$ Total	Student	$ Total	Student	$ Total
Raj	$8.59	Raj	$22.32	Raj	$6.74
Jada	$4.33	Jada	$3.67	Jada	$14.82
Estelle	$2.56	Estelle	$3.19	Estelle	$17.72
Alex	$4.18	Alex	$9.55	Alex	$2.89

Student	$ Total
Raj	$37.65
Jada	$22.82
Estelle	$23.47
Alex	$16.62

Method 1 (Rounded)
$37.70
$22.80
$23.50
$16.60

Lesson 2: Leftovers
Divvy It Up! (pages 51–52)

Number of Pieces in Bag	Pieces per Student	Pieces for Mr. Kuo	Number of Pieces in Bag	Pieces per Student	Pieces for Mr. Kuo
49	6	1	57	7	1
50	6	2	58	7	2
51	6	3	59	7	3
52	6	4	60	7	4
53	6	5	61	7	5
54	6	6	62	7	6
55	6	7	63	7	7
56	7	0	64	8	0

1. 8; because you are dividing the candy among 8 students in the class.
2. Number of Pieces in Bag
3. Pieces per Student
4. Pieces for Mr. Kuo
5. 7
6. 0
7. Yes
8. Yes
9. Answers will vary.
10. 0, 1, 2, 3, 4, 5, 6, 7

Answer Key (cont.)

By the Numbers (page 53)

1.
   a. 7 pieces
   b. 0 through 14
   c. Yes
   d. 112
2.
   a. 8 pieces
   b. 0, 1, or 2
   c. Now
3. 5 pieces
4. Mr. Kuo would receive 4, 5, 6, 7, or 8 pieces.
5. When the rounded quotient is 5, Mr. Kuo would receive 4 or 5 pieces. When the rounded quotient is 6, Mr. Kuo would receive 6, 7, or 8 pieces.

Lesson 3: Half Again

It All Adds Up (pages 57–58)

1. $\frac{1}{4}$	2. $\frac{2}{8}$	3. $\frac{3}{12}$
4. $\frac{4}{16}$	5. $\frac{5}{20}$	6. $\frac{6}{24}$
7. $\frac{7}{28}$	8. $\frac{8}{32}$	9. $\frac{9}{32}$
10. $\frac{10}{40}$	11. $\frac{20}{80}$	12. $\frac{19}{80}$

1. $\frac{1}{6}$	2. $\frac{2}{10}$	3. 0
4. $\frac{2}{12}$	5. $\frac{2}{6}$	6. $\frac{4}{14}$
7. $\frac{4}{18}$	8. $\frac{3}{20}$	9. $\frac{8}{24}$
10. $\frac{7}{16}$	11. $\frac{1}{50}$	12. $\frac{1}{200}$

1. The numerator is always one half of the denominator.
2. It simplified the fraction to an equivalent fraction with a smaller numerator and denominator.

It All Adds Up (pages 57–58) (cont.)

3. The numerator of the new fraction is found by taking half of the denominator and then subtracting the first numerator from that value. The denominator of the new fraction is the same as the denominator of the second fraction. Then I subtracted the addend from $\frac{1}{2}$ to find the missing addend.

Different Halves (page 59)

1. 1
2. $\frac{2}{10}$
3. $\frac{1}{5}$ is simplified from $\frac{2}{10}$; $\frac{2}{10}$ is equivalent to $\frac{1}{5}$.
4. It is the simplified form. Multiplying by a form of 1 does not change the value of the number.
5. $\frac{3}{10}$
6. Answers will vary.
7. $\frac{6}{14}$
8. $\frac{6}{14}$
9. Answers will vary.
10. $\frac{5}{26}$; answers will vary.

Lesson 4: More or Less

Fair Prices (pages 63–64)

1.
   a. $\frac{2}{5}$
   b. $\frac{3}{7}$
   c. $\frac{14}{35}$
   d. It is equal.
   e. $\frac{15}{35}$
   f. They are equal.
   g. Cherry pie; 3 pieces of the pie that is cut into 7 pieces or 3/7.
   h. $\frac{2}{5} < \frac{3}{7}$ and $\frac{3}{7} > \frac{2}{5}$
2.
   a. $\frac{6}{7}$
   b. $\frac{7}{9}$
   c. $\frac{54}{63}$
   d. It is equal.
   e. $\frac{49}{63}$

Answer Key (cont.)

Fair Prices (pages 63–64) (cont.)

f. It is equal to 1 and 7 is the denominator of the first fraction.

g. $\frac{5}{7} < \frac{7}{9}$ and $\frac{7}{9} > \frac{5}{7}$

h. The lemon cake; 7 pieces from cake cut into 9 pieces, or $\frac{7}{9}$

Who Has More? (page 65)

1.
a. $\frac{18}{18}$
   $\frac{90}{216}$

b. $\frac{12}{12}$
   $\frac{84}{216}$

c. Elena

d. Jill

e. $\frac{5}{12} > \frac{7}{18}$ and $\frac{5}{18} < \frac{7}{12}$

f. $\frac{4}{13} < \frac{9}{25}, \frac{9}{25} > \frac{4}{13}$

g. Answers will vary.

2.
a. $\frac{25}{25}$

b. $\frac{13}{13}$

c. $\frac{9}{25}$

d. Jill

e. Answers will vary.

f. Answers will vary.

Lesson 5: Big Buck$

Uncle Buck$ (pages 69–70)

Age	$$$
0	1
1	2
2	4
3	8
4	16
5	32
6	64
7	128
8	256

a. 1 × 2; 1

b. 2 × 2; 2

c. 2 × 2 × 2; 3

d. 2 × 2 × 2 × 2 × 2; 8

e. 2 × 2 × 2 × 2 × 2 × 2 × 2 × 2 × 2 × 2 × 2 × 2 × 2 × 2 × 2 × 2 × 2 × 2 × 2 × 2; 20

Age	$$$	Age	$$$	Age	$$$
0	1	6	729	12	531,441
1	3	7	2,187	13	1,594,323
2	9	8	6,561	14	4,782,969
3	27	9	19,683	15	14,348,907
4	81	10	59,049	16	43,046,721
5	243	11	177,147	17	129,140,163

a. 1 × 3; 1

b. 3 × 3; 2

c. 3 × 3 × 3; 3

d. 3 × 3 × 3 × 3 × 3 × 3 × 3; 3

e. 3 × 3 × 3 × 3 × 3 × 3 × 3 × 3 × 3 × 3 × 3 × 3 × 3 × 3 × 3 × 3; 16

f. Many students will immediately believe that Hannah's aunt is more generous because her last gift is larger. Students may suggest that you need to look at the total of all the gifts. Hannah's Uncle's total is $134,217,721 and her aunt's total is $193,710,244, so her aunt's total is larger.

Dollars and Days (page 71)

1.

Day	Factors	Expression	Base	Exponent	Value
1	7	7^1	7	1	7
2	7 × 7	7^2	7	2	49
3	7 × 7 × 7	7^3	7	3	343
4	7 × 7 × 7 × 7	7^4	7	4	2,401
5	7 × 7 × 7 × 7 × 7	7^5	7	5	16,807
6	7 × 7 × 7 × 7 × 7 × 7	7^6	7	6	117,649
7	7 × 7 × 7 × 7 × 7 × 7 × 7	7^7	7	7	823,543
8	7 × 7 × 7 × 7 × 7 × 7 × 7 × 7	7^8	7	8	5,764,801

2.

Day	Factors	Expression	Base	Exponent	Value
1	8	8^1	8	1	8
2	8 × 8	8^2	8	2	64
3	8 × 8 × 8	8^3	8	3	512
4	8 × 8 × 8 × 8	8^4	8	4	4,096
5	8 × 8 × 8 × 8 × 8	8^5	8	5	32,768
6	8 × 8 × 8 × 8 × 8 × 8	8^6	8	6	262,144
7	8 × 8 × 8 × 8 × 8 × 8 × 8	8^7	8	7	2,097,152
8	8 × 8 × 8 × 8 × 8 × 8 × 8 × 8	8^8	8	8	16,777,216

Age	$$$	Age	$$$
9	512	18	262,144
10	1,024	19	524,288
11	2,048	20	1,048,576
12	4,096	21	2,097,152
13	8,192	22	4,194,304
14	16,384	23	8,388,608
15	32,762	24	16,777,216
16	65,536	25	33,554,432
17	131,072	26	67,108,864

Answer Key (cont.)

Power Lesson 1: Favorites
Group Response Sheet (page 77)
Answers will vary according to student responses.

Tally Sheet (page 78)
Answers will vary according to student responses.

Fractions and Decimals (pages 79–80)
Answers will vary according to student responses.

The Rest of the Story (page 81)
Answers will vary according to student responses.

Lesson 6: How Many Cookies?
Sugar Cookies (pages 85–86)

1.
   a. 96 cookies
   b. 8 dozen
   c. 2 batches
   d. 5 cups flour
   e. 2 teaspoons baking soda
   f. 1 teaspoon baking powder
   g. 2 cups butter
   h. 3 cups sugar
   i. 2 teaspoons vanilla
   j. 2 eggs

2.
   a. 216 cookies
   b. 18 dozens
   c. $4\frac{1}{2}$ batches
   d. $11\frac{1}{4}$ cups flour
   e. $4\frac{1}{2}$ teaspoons baking soda
   f. $2\frac{1}{4}$ teaspoons baking powder
   g. $4\frac{1}{2}$ cups butter
   h. $6\frac{3}{4}$ cups sugar
   i. $4\frac{1}{2}$ teaspoons vanilla
   j. $4\frac{1}{2}$ eggs; egg comments will vary.

Lots of Cookies (page 87)

1. flour: 40 cups; sugar: 8 cups; butter: 16 cups; vanilla: 8 tablespoons; milk: 16 tablespoons
2. 32 dozen
3. $\frac{1}{8}$
4.
   5 cups flour
   1 cup sugar
   2 cups butter
   2 tablespoons milk
   1 tablespoon vanilla extract

Lesson 7: Fill in the Blank
Missing Numbers (pages 91–92)
Answers will vary depending on the problems given to them by the TI-15.

Equation	? (Solution)	How
305 – 5 = ?	300	Subtracted 5 from 305.
180 – ? = 80	100	Subtracted 80 from 180.
? – 120 = 10	130	Added 120 and 10.
161 – 11 = ?	150	Subtracted 11 from 161.
190 – 17 = ?	173	Subtracted 17 from 190.
? – 88 = 120	208	Added 88 and 120.

Write You Own (page 93)

1. Equations will vary; subtract.
2. Equations will vary; subtract.
3. Equations will vary; add.
   Word problems wil vary.

Lesson 8: Moving Along
Snail Mail (pages 97–98)

1.
   a. 30 mm
   b. 15 mm

Day	0	1	2	3	4	5	6	7	8	9	10
Position (in mm)	42	57	72	87	102	117	132	147	162	177	192

   c. 1 and 57
   d. 2 and 72
   e. 3 and 87

2.
   a. 23 mm
   b. The second snail moved faster.
   c. Op1 + 2 3 Op1 To add 23 mm each day.

Day	0	1	2	3	4	5	6	7	8	9	10
Position (in mm)	37	60	83	106	129	152	175	198	221	244	267

3.
   a. 27 mm
   b. add 27
   c. 23 mm
   d. add 23

Answer Key (cont.)

Snail Mail (pages 97–98) (cont.)

e.

Day	0	1	2	3	4	5	6	7	8	9	10
Gray Position (in mm)	53	80	107	134	161	188	215	242	269	296	323
Brown Position (in mm)	75	98	121	144	167	190	213	236	259	282	305

f. Days 5 and 6

Hikers (page 99)

1.
a. 34 miles

b.

Day	0	1	2	3	4	5	6	7	8
Position (miles)	39	73	107	141	175	209	243	277	311

c. 39 miles
d. Days 0–4: [Op] – 3 [4] [Op] 1 [0] 7 [Op] etc. Days 7–8: [Op] + 3 [4] [Op] 2 [4] 3 [Op] etc.
e. Find the total distance traveled (311 – 39 = 272) and then divide by the number of days of travel (272 ÷ 8 = 34).

2.
a. and c.

Day	0	1	2	3	4	5
Position (miles)	50	82	114	146	178	210

b. 32 miles
d. Start with 50 and add 32 miles every day.

Lesson 9: Order Matters

Which Way Did They Go? (pages 103–104)

1. multiplication
2. multiplication
3. division
4. multiplication
5. Multiplication, Division, Addition, Subtraction
6. 10
7. 12
8. 40

Which Way Did They Go? (pages 103–104) (cont.)

9. 38
10. 43
11. 45
12. 5
13. 8
14. 4
15. 11
16. 15
17. 4
18. 8
19. 20
20. 5
21. 40
22. 32
23. 8
24. 5
25. 200

All Mixed Up (page 105)

1. 21
2. 50
3. 14
4. 4
5. 5
6. 2
7. 20
8. 6
9. 11
10. 8
11. 43
12. 4
13. 4
14. 14
15. 15

Lesson 10: The Shapes of Numbers

Dots (pages 109–110)

1. Term: 1, 2, 3, 4; Value: 1, 3, 6, 10
   a. 1
   b. 2
   c. 3
   d. 4
   e.

2.

Rows	1	2	3	4	5	6	7	8	9
Dots	1	3	6	10	15	21	28	36	45

3. Term: 1, 2, 3, 4, 5; Value: 1, 4, 9, 16, 25
   a. 3
   b. 5
   c. 7
   d. 9

Answer Key (cont.)

Dots (pages 109–110) (cont.)

4.

Term	1	2	3	4	5	6	7	8	9
Value	1	4	9	16	25	36	49	64	81

The Italian Connection (page 111)

Term	1	2	3	4	5	6	7	8	9
Value	1	2	3	5	8	13	21	34	

Answers will vary, but each number in the sequence needs to be the sum of the two numbers before it.

Power Lesson 2: A New Look

Dozens and Dozens (page 117–118)

1.

Number of cookies	1	2	3	4	5	6	7	8	9
Earnings	.50	1.00	1.50	2.00	2.50	3.00	3.50	4.00	4.50

Serena's Oatmeal Cookies graph

a. They increase by 50 cents.
b. The graph goes up by the same amount for each cookie.
c. $6.50
d. 0

Dozens and Dozens (page 117–118) (cont.)

2.

Dozens of cookies	1	2	3	4	5	6
Earnings	5	10	15	20	25	30

Fritz's Chocolate Chip Cookies graph

The Cookie Business (pages 119–121)

1.

Month	Jun.	Jul.	Aug.	Sept.	Oct.	Nov.	Dec.	Jan.	Feb.
Dozens of cookies	50	100	150	200	250	300	350	400	450
Expenses in Dollars	400	500	600	700	800	900	1,000	1,100	1,200

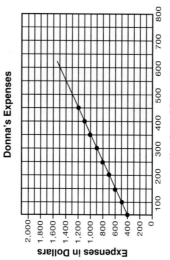

Donna's Expenses

a. $100 each month
b. $1,700.00
c. When no cookies are sold, the expenses are $300.00.

2.
a. $2.50 per dozen
b. $950.00 profit; 500 × $2.50 = $1,250; $1,250 − 300 = $950.00
c. Use Op1 to multiply by 2.50 and Op2 to subtract 300.

Answer Key (cont.)

The Cookie Business (pages 119–121) (cont.)

3.

Month	June	July	Aug.	Sept.	Oct.	Nov.	Dec.	Jan.	Feb.
Dozens of cookies	50	100	150	200	250	300	350	400	450
Profit	-$175	-$50	$75	$200	$325	$450	$575	$700	$825

a. She is losing money.
b. She starts to make a profit.
c. $0.00
d. 120 dozen

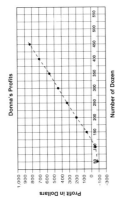

Lesson 11: Mini Me

Scaled Down (pages 125–126)

Individual measurements will vary.

1. Dividing by 4 is the same as multiplying by $\frac{1}{4}$.
2. Press [×] [1] [n] [d] [8] [d] and [Enter], or press [8] and [Enter].
3. The legs are not straight down.
4. The length of the head, the neck, the trunk, and the legs.
5. Answers depend on student measurements.

Scaled Up (page 127)

1. Answers will vary, but should include multiplying the scale drawing measurement by 4 or dividing the scale drawing measurement by $\frac{1}{4}$.
2. Answers will vary.
3. Answers will vary, but are likely to include estimation that is required when measuring.

Lesson 12: On the Ball

Big and Small (pages 131–132)

1. Yes
2. smaller
3. larger
4. Answers will vary, but the true value is about 1.5.
5–7. Characteristics will vary, but calculated ratios should be close to 0.68.
8. No, because measurement with a ruler is imprecise.
9. Answers will vary, but should be close to 29.41 mm.
10. Answers will vary, but should be close to 13.60 mm.

Another Look (pages 133)

1. Reduction
2. Answers will vary, but the ratio is approximately 1.21.
3. Values on the chart will vary depending on parts chosen to measure.
4. Answers will vary.

Lesson 13: It's on the Map

Mapping Measurement Mountain (pages 137–138)

1. The routes and distances will vary according to student choices.
2. 25 miles
3. Answers will vary according to the chosen route.
4. 40 kilometers
5. Answers will vary according to the chosen route.

My Math Map (page 139)

1. Answers will vary according to the scale chosen.
3.
  a. 68 miles
  b. 93 miles
4. About 121 miles
5. About 55 miles
6. It is faster to take Highway 2^2 because it is a shorter distance.

Lesson 14: Scale that Structure

Around the World (page 143)

	Taj Mahal	The White House	Arc de Triomphe	Buckingham Palace
Height	42.6 in.	14 in.	33 in.	15.8 in.
Length	37.2 in.	33.6 in.	29.6 in.	70.8 in.

My Structure (pages 144–145)

Answers will vary according to student measurements.

Lesson 15: Painting Flags

Similar and Congruent (page 149)

1. They both have the same size and shape.
2. They have the same shape, but not necessarily the same size.
3. They are similar. Both flags are rectangles, but they are not the same size.
4. Flag A: 12 units x 22 units; Flag B: 24 units x 22 units

© Shell Education #50616—30 Mathematics Lessons Using the TI-15 **249**

Answer Key (cont.)

Similar and Congruent (page 149) (cont.)
5. Flag A—22:12 or 1:8; Flag B—22:24 or .92
6. The ratios are different because the shapes are not congruent.
7. Flag A: 264 sq. units; Flag B: 528 sq. units
8. Drawings will vary.
9. Drawings and lengths will vary.

Stripes (pages 150–151)

Long Stripes

Width	Length	Area
1 in.	24.7 in	24.7 sq. in.
2 in.	49.4 in	98.8 sq. in.
3 in.	74.1 in	222.3 sq. in.
4 in.	98.8 in	395.2 sq. in.

Short Stripes

Width	Length	Area
1 in.	14.82 in	14.82 sq. in.
2 in.	29.64 in	59.28 sq. in.
3 in.	44.46 in	133.38 sq. in.
4 in.	59.28 in	237.12 sq. in.

1. 4
2. 9
3. 16
4. yes
5. 100
6. 1 square foot
7. 144 square inches
8. 144 square inches
9. 400 × 144 = 57,600 sq. in.

Power Lesson 3: Shape Investigations

Shape Scavenger Hunt (pages 157–158)
Answers will vary depending on the shapes chosen by students.

Investigating Shapes (page 159)
Answers will vary depending on the mystery shape chosen by each student.

Irregular Shapes (page 160)
1. perimeter: 158cm; area: 1,118 cm2
2. perimeter: 130 m; area: 780 m2
3. perimeter: 140 in.; area: 784 in.2

Designing a Park (page 161)
Designs will vary, but students must include every item on the checklist.

Lesson 16: Hamster Haven

Harvey the Hamster (page 165–166)
1. a circle
2. the distance around the circle (the circumference)
3. circumference
4. π times the diameter
5. about 12.57 cm
6. about 351.96 sq. cm
7. Student measurements may vary due to duplication.

Diameter	Circumference	Area
4.5 cm	14.14 cm.	395.92 sq. cm
5 cm	15.71 cm.	439.88 sq. cm
5.5 cm	17.28 cm.	483.84 sq. cm
6 cm	18.35 cm.	527.79 sq. cm

8. Answers will vary.
9. Answers will vary depending on the size of the circle chosen.
10. Answers will vary depending on the size of the circle chosen.
11. Answers will vary depending on the size of the circle chosen.
12. Models will vary.

Diameter	Circumference	Area
6.5 cm	20.42 cm.	571.76 sq. cm
7 cm	21.99 cm.	616.00 sq. cm
7.5 cm	23.56 cm.	659.68 sq. cm
8 cm	25.13 cm.	703.64 sq. cm

Four Walls (page 167)
1. Answers will vary, but the side of the square should equal the diameter of the circle.
2. rectangle
3. 28 cm
4. equal to the side of the square
5. Answers will vary.
6. 4
7. Four times the answer in #5.
8. Check student sketches.
9. Add one to the answer to #4.
10. Multiply the answer to #9 times 4.
11. the box

Answer Key (cont.)

Lesson 17: Measuring Mania

Use Your Feet! (page 171)
1. Answers will vary, but most members of the group will likely have different-sized feet, resulting in different answers.
2. He would get a smaller number because his feet are larger.

Measure That Rectangle (pages 172–173)
1. length: 5 cm; width: 2.5 cm; perimeter: 15 cm; area: 12.5 cm²; Finger measurements will vary.
2. length: 8 cm; width: 6 cm; perimeter: 28 cm; area: 48 cm²; Finger measurements will vary.
3. length: 6 cm; width: 5 cm; perimeter: 22 cm; area: 38 cm²; Finger measurements will vary.
4. length: 3.5 cm; width: 1 cm; perimeter: 9 cm; area: 3.5 cm²; Finger measurements will vary.
5–7. Answers will vary.

Lesson 18: Cover It Up

Beneath the Feet (pages 177–178)
1. 140 sq. ft.
2. $1,976.80
3. $1,216.60
4. cost and installation ($2.52 and $3.76)
5. 140 × ($2.52 + 3.76)
6. $879.20
7. Answers will vary.
8. The dimensions of Kimi's room are not whole numbers.
9. 12 inches
10. $10 \frac{3}{12}$
11. $12 \frac{6}{12}$
12. $128 \frac{8}{144}$
13. $1,809.13
14. $1,113.41
15. $804.63
16. Answers will vary.

Kitchen Floor (page 179)
1. Answers will vary.
2. 192 square feet
3. 27,648 square inches
4. 36 square inches
5. 768 tiles
6. $1,774.08
7. $1,741.44
8. $3,515.52
9. 144 square inches
10. 192 tiles
11. $837.12
12. $1,741.44
13. $2,578.56
14. Tile #2

Lesson 19: Investigating Volume

Boxes of Volume (page 183)
Answers will vary depending on boxes provided.

Build It! (pages 184–185)
Answers will vary.

Lesson 20: Fish for the Science Fair

Fish Frenzy (page 189)
1.
   a. 128 oz.
   b. 256 oz.
2. fish; ounces
3.
   a. 1 gallon
   b. 2 gallons
4. ounces; gallons
5. Tank 1: 29 gallons; Tank 2: 28 gallons

Fish Food (pages 190–191)
1. $\frac{1 \text{ cup}}{8 \text{ oz.}} \times \frac{4 \text{ oz.}}{1} = \frac{1}{2}$ cup wheat germ

   $\frac{1 \text{ cup}}{8 \text{ oz.}} \times \frac{6 \text{ oz.}}{1} = \frac{3}{4}$ cups corn flour

   $\frac{1 \text{ cup}}{8 \text{ oz.}} \times \frac{6 \text{ oz.}}{1} = \frac{3}{4}$ cups ground barley

2. $\frac{3 \text{ teaspoons}}{1 \text{ tablespoon}} \times \frac{2 \text{ tablespoons}}{1} = 6$ teaspoons spinach powder

3. $\frac{1 \text{ cup}}{48 \text{ teaspoons}} \times \frac{6 \text{ teaspoons}}{1} = \frac{1}{8}$ cup spinach powder

4.
   a. 32 oz.
   b. 32 oz.; 4 cups
   c. 48 oz.
   d. 48 oz.; 6 cups
   e. 48 oz.
   f. 48 oz.; 6 cups
   g. 8 oz.
   h. 8 oz.; 1 cup

Answer Key (cont.)

Fish Food (pages 190–191) (cont.)

5.
  a. $4\frac{1}{2}$; 8 jars
  b. $6\frac{3}{4}$; 8 jars
  c. $6\frac{3}{4}$; 8 jars
  d. $1\frac{1}{8}$; 8 jars
  e. 8 jars

Power Lesson 4: Easy as Pi

Parts of a Circle (page 197)

radius/diameter	diameter/circumference	chord/diameter
Alike: both are a line segment; both are positioned in the center of the circle; both are a measure of distance or length	**Alike:** both are a measure of distance	**Alike:** both are a measure of distance; both are a line segment
Different: diameter crosses through the circle; radius only goes to the center of the circle	**Different:** diameter is the width of the circle; circumference is the distance around the circle	**Different:** diameter crosses through the center of the circle
Relationship: diameter is twice the length of the radius	**Relationship:** the ratio of diameter to circumference is always the same, 3.14 or 22/7 (pi)	**Relationship:** the diameter could also be considered a chord since the diameter, like a chord, joins two points on a circle

Round About (page 198)

1. circumference
2. diameter
3. Answers will vary.
4. Answers will vary.
5. Answers will vary.
6. yes

Close Enough? (pages 199–201)

1. 3 cm
2. 6 cm
3. Student measurements may vary due to duplication.

Figure	Length of side	Perimeter	Ratio
Triangle	4.9 cm	14.7 cm	2.45
Square	4 cm	16 cm	2.67
Pentagon	3.3 cm	16.5 cm	2.75
Hexagon	2.9 cm	17.4 cm	2.9
Octagon	2.2 cm	17.6 cm	2.93
17-gon	1 cm	18.7 cm	3.11

4.
  a. triangle
  b. 17-gon
  c. 17-gon
  d. The number of sides would increase.
  e. A circle because the angles would be larger and larger and the side lengths would get smaller and smaller.
  f. circumference
  g. It gets larger and closer to pi.
5. π
6. 3.141592654
7. 18.49 cm
8. Answers will vary.

Lesson 21: In the Long Run

Heads or Tails (pages 205–206)

1. Answers will vary. Most students will predict 10 for both heads and tails.
2. Answers will vary. Most students will predict $\frac{10}{20}$ or $\frac{1}{2}$.
3. Answers will vary. Most students will get $\frac{1}{2}$.
4. Answers will vary. Most students will predict 10/20 or $\frac{1}{2}$.
5. Answers will vary. Most students will get $\frac{1}{2}$.
6. Results will vary.
7. 20
8. Answers will vary.
9. $\frac{1}{2}$
10–14. Answers depend on the class results.
15. The class results should be the closest to $\frac{1}{2}$ because the number of trials is larger.

Answer Key (cont.)

Fair Play (page 207)

1. 6
2. Answers will vary. Most students will predict $\frac{1}{6}$ of the 60 times; or 10 times.
3. Results will vary.
4. Results will vary.
5. 60
6–8. Answers will depend on the class results.

Lesson 22: Bowling For Dollars
Rolling Along (pages 211–212)

1–2.

Red Team	Total	Average
Julia	315	105
Jorge	306	102
Nadia	306	102
Jeffrey	450	150
Blue Team	**Total**	**Average**
Barry	324	108
Serena	408	136
Raymond	171	57
Anna	306	102

3. Jorge, Nadia, and Anna; They all have the same total points.
4. Blue: 100.75; Red: 114.75; add together all averages for each team and then divide by 4
5. lowest score = 1,378; team average = 344.5
6. lowest score = 1,210; team average = 302.5
7. 194 points
8. 927 points

9.

Purple Team	Game 1	Game 2	Total after Game 2	Game 3	Total	Average
Joe	86	105	191	varies	varies	varies
Tamryn	128	129	257	194	451	150.33
Mika	100	150	250	varies	varies	varies
Eugene	88	120	208	varies	varies	varies

10. $302.50

On a Roll (page 213)

1. 143 points
2. 564 points
3. 129 points
4.

Mrs. Howard's Team	Game 1	Game 2	Average	Mr. Garcia's Team	Game 1	Game 2	Average
Evelyn	143	133	138	Faraah	147	111	129
Hans	143	157	150	Ajay	138	144	141
Bianca	143	101	122	Gracie	150	166	158
Ernest	143	149	146	Ziva	129	91	110

5. 139 points
6. 134.5 points
7. Mrs. Howard's team

Lesson 23: It's In the Bag
Bag of Chips (pages 217–218)

Answers will vary according to chips students will receive.

Mystery Chips (page 219)

Answers will vary according to chips students will receive.

Lesson 24: They Add Up
Let's Get Rolling (pages 223–224)

1–6. Answers will vary depending on individual results.
7. 50 x the number of students in the class
8–16. Answers will vary depending on class results.

Toss Those Cubes! (page 225)

1.

1st Cube	2nd Cube	2nd Cube	2nd Cube	2nd Cube	2nd Cube	2nd Cube
1	1	2	3	4	5	6
2	2	3	4	5	6	1
3	3	4	5	6	1	2
4	4	5	6	1	2	3
5	5	6	1	2	3	4
6	6	1	2	3	4	5

36 combinations

2. 36 combinations

Answer Key (cont.)

Toss Those Cubes! (page 225) (cont.)

3.

Sum	2	3	4	5	6	7	8	9	10	11	12
Combinations	1+1	1+2 2+1	1+3 2+2 3+1	1+4 2+3 3+2 4+1	1+5 2+4 3+3 4+2 5+1	1+6 2+5 3+4 4+3 5+2 6+1	2+6 3+5 4+4 5+3 6+2	3+6 4+5 5+4 6+3	4+6 5+5 6+4	5+6 6+5	6+6
Fraction	1/36	1/18	1/12	1/9	5/36	1/6	5/36	1/9	1/12	1/18	1/36
Percent	2.8 %	5.6 %	8.3 %	11.1 %	13.9 %	16.7 %	13.9 %	11.1 %	8.3 %	5.6 %	2.8 %

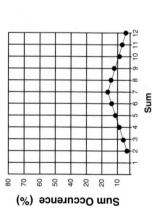

Number Cube Sums

4. The graphs should be similar. It is easy to see which is larger and how much the difference is.

Lesson 25: Typically
Birthday Weather (pages 229–230)

1. 23 temperatures
2. Subtraction is the difference between the years. The number of temperatures must include data from both those years, so it is 1 more.
3. 38.3° F
4. 64° F
5. 7° F
6. 25.7° F
7. 31.3° F
8. The temperatures are not evenly spread out in the list.
9. Answers will vary. Students may think they are the same or one is larger than the other.
10. 7°, 16°, 17°, 28°, 28°, 29°, 30°, 30°, 32°, 34°, 35°, 37°, 41°, 42°, 45°, 45°, 50°, 52°, 57°, 59°, 59°, 64°
11. yes, 37° F

Birthday Weather (pages 229–230) (cont.)

12. 27° F
13. 30° F
14. no; The median is the middle value in the set of data, not the middle number between the highest and lowest number in the set.
15. larger; The extreme temperatures make the mean smaller than the median.
16. yes; 45°
17. Answers will vary. Students should think that the typical temperature is around the median of 37°.
18. Answers will vary.

Birthday Game (page 231)

1. The heights are listed in feet and inches.
2. Multiply the number of feet by 12 and then add the number of inches.
3. 48 in.
4. 48, 50, 50, 51, 53, 54, 55, 55, 57, 57, 57, 57, 58, 59, 59, 63
5. 55.18
6. 56
7. 57
8. Answers will vary.

Power Lesson 5: Graph It!
Frozen Flavors (pages 237–239)

1. chocolate, peanut butter, cake batter, strawberry, vanilla
2. Answers will vary.
3. Answers will vary.
4. vanilla: 1 student; chocolate: 12 students; strawberry: 4 students; peanut butter: 8 students; cake batter: 5 students
5. 12 more students
6. 30 students
7. 60 students
8. The total can be found by adding the number of responses in each category.
9. 2 more students
10. Mr. Ferraro's class
11. Mrs. Wheeler's class
12. chocolate, cake batter, peanut butter, and strawberry
13. Yes, because their votes would make vanilla have one more vote than strawberry.

Answer Key (cont.)

A Flavorful Dessert (page 240)

Answers depend on the class data.

Favorite Flavors (page 241)

	Vanilla	Chocolate	Strawberry	Peanut Butter	Cake Batter
Number of People	3	70	7	41	20

Frozen Yogurt Palace Favorite Flavors

Top That! (page 242)

Answers depend on the class data.

Appendix B

Teacher Resource CD Contents

Folder: Student Reproducibles

Pg.	Title	Filename
45	Round It Up	page045.pdf
47	More Money	page047.pdf
51	Divvy It Up!	page051.pdf
53	By the Numbers	page053.pdf
57	It All Adds Up	page057.pdf
59	Different Halves	page059.pdf
63	Fair Prices	page063.pdf
65	Who Has More?	page065.pdf
69	Uncle Buck$	page069.pdf
71	Dollars and Days	page071.pdf
77	Group Response Sheet	page077.pdf
78	Tally Sheet	page078.pdf
79	Fractions and Decimals	page079.pdf
81	The Rest of the Story	page081.pdf
85	Sugar Cookies	page085.pdf
87	Lots of Cookies	page087.pdf
91	Missing Numbers	page091.pdf
93	Write Your Own	page093.pdf
97	Snail Mail	page097.pdf
99	Hikers	page099.pdf
103	Which Way Did They Go?	page103.pdf
105	All Mixed Up	page105.pdf
109	Dots	page109.pdf
111	The Italian Connection	page111.pdf
117	Dozens and Dozens	page117.pdf
119	The Cookie Business	page119.pdf
125	Scaled Down	page125.pdf
127	Scaled Up	page127.pdf
131	Big and Small	page131.pdf
133	Another Look	page133.pdf
137	Mapping Measurement Mountain	page137.pdf
139	My Math Map	page139.pdf
143	Around the World	page143.pdf
144	My Structure	page144.pdf
149	Similar and Congruent	page149.pdf
150	Stripes	page150.pdf
157	Shape Scavenger Hunt	page157.pdf
159	Investigating Shapes	page159.pdf
160	Measuring Irregular Shapes	page160.pdf
161	Designing a Park	page161.pdf
165	Harvey the Hamster	page165.pdf
167	Four Walls	page167.pdf
171	Use Your Feet!	page171.pdf
172	Measure That Rectangle	page172.pdf
177	Beneath the Feet	page177.pdf
179	Kitchen Floor	page179.pdf
183	Boxes of Volume	page183.pdf
184	Build It!	page184.pdf
189	Fish Frenzy	page189.pdf
190	Fish Food	page190.pdf

Folder: Student Reproducibles (cont.)

Pg.	Title	Filename
197	Parts of a Circle	page197.pdf
198	Round About	page198.pdf
199	Close Enough?	page199.pdf
205	Heads or Tails	page205.pdf
207	Fair Play	page207.pdf
211	Rolling Along	page211.pdf
213	On a Roll	page213.pdf
217	Bag of Chips	page217.pdf
219	Mystery Chips	page219.pdf
223	Let's Get Rolling	page223.pdf
225	Toss Those Cubes!	page225.pdf
229	Birthday Weather	page229.pdf
231	Birthday Game	page231.pdf
237	Frozen Flavors	page237.pdf
240	A Flavorful Dessert	page240.pdf
241	Favorite Flavors	page241.pdf
242	Top That!	page242.pdf

Folder: Lesson Support Files

Pg.	Title	Filename
n/a	Student-Friendly Glossary	glossary.pdf
140	Picture of Taj Mahal	mahal.jpg
140	Picture of White House	house.jpg
140	Picture of Arc de Triomphe	arc.jpg
140	Picture of Buckingham Palace	palace.jpg
146	Flags from Around the World	flags.jpg
146	Flag of Greece	greece.jpg

Folder: Teacher Resource Files

Pg.	Title	Filename
20	Frayer Model Template	frayer.doc
21	Word Wizard Chart	wizard.doc
33	Check-Off List	checkofflist.doc
34	Damage Report	damage.doc
36	Parent Letter	letter.doc
38	Completion Grades Template	grades.doc
39	General Rubric	genrubric.doc
40	Create Your Own Rubric	rubric.doc